The Exclusive Guidebook to the WORLD of MEDICINE

How to Empower Yourself and Become More Informed by Learning the Inside Secrets of the Medical System

By
Dana Jacoby

ISBN 978-0-6151-3874-9

DEDICATION:

This book is dedicated to all patients, students, and practitioners of medicine. My hope is that it will serve as a guide through the confusion of what I have come to term the "medical world." May everyone who reads it gain a better understanding of the medical world and develop the ability to communicate better with public servants in the healthcare realm.

CONTENTS:

INTRODUCTION:

The idea for this book was originally developed as I sat on the shores of Florida while visiting my grandparents in April of 2002. At the time, I was employed as a pharmaceutical sales representative for Ortho-McNeil Pharmaceutical, a medical division of Johnson and Johnson.

My daily job consisted of talking about medicines to neurologists, family practice physicians, internal medicine doctors, the Emergency Department, orthopedic surgeons, pain management specialists, and many other physician specialties throughout hospitals and medical clinics in my territory in St. Louis, MO.

Since that time, I have continued on in positions in the medical industry in sales and management, while earning a master of business management certification as well as a master of health systems degree. The dream to write this book has stuck with me since that day in April of 2002.

With the above paragraph as an introduction, here is the brief history that led me up to the actual writing of this book…

Although I still believe that we live in a country with the best medical system in the world, I have witnessed some flaws with the medical system during my days as a medical and pharmaceutical representative. This has made me passionate about changing the system. Throughout my career in healthcare it has often bothered me to see and hear about packed patient waiting rooms, lack of patient education, and patients who received less than 15 minutes of face time while seeking high quality care.

My days as a medical sales rep were initially filled with medical dinners, glossy sales aids, and numerous free drug samples that were worth thousands of dollars. Through personal relationships, speaking engagements, and pharmacy calls, I was a successful rep in multiple medical markets with a good long-term career potential. I enjoyed the time spent with the physicians that I called on and respected their knowledge immensely. They had worked long, hard hours to earn specialties in their respective fields, and I was in awe at their abilities and dedication to the field of medicine.

Still, at times, something in my heart just did not feel right when looking at the overall medical system. I had grown up in a family with a mother who was a patient-focused physical therapist. Mom had gone into medicine originally to care for the patient lovingly and willingly. Additionally, my maternal grandfather had been a family practice general practitioner who had dedicated his life to serving others through medicine. As a result I was raised to believe that patient care and education was the reason for the medical system's existence.

Based on this upbringing, it just did not seem right for me to cater lunches, dinners, and give thousands of dollars in honoraria in order to ask a physician to give a patient a certain drug without adding patient education into the mix.

I constantly wanted to do more for patient education every time I entered a physician's office. As a sales rep I understood the service that I provided, but felt that there was a lack of true patient education for navigating through the pharmaceutical and medical environment.

I justified to myself on a continual basis that my sales job paid well, and I was content not only with my job, but also the medical clinics and hospitals that I was working with.

Then in 2002, while I was visiting my grandparents on Marco Island, FL, my perspective was changed forever.

In the winter of 2001 my grandfather had been diagnosed with prostrate cancer and a neurological condition called myasthenia gravis. Although his health was still good, it was tough for me to see a man who had been in tip-top tennis playing condition two months prior be anything less than perfect in the area of health.

As I listened patiently to his stories of visiting physician after physician to seek a diagnosis and treatment, it became apparent to me that the crowded waiting rooms and lack of patient education that I had witnessed as a rep were beginning to hit a bit closer to home.

Prior to retirement my grandfather had been a chief systems engineer for Lockheed Martin. He had been trained as a radar technologist in WW II, and then had quickly risen through the ranks at Lockheed to become one of the key top classified engineers in the country. At the time of his retirement he could still cite numerous shuttle launches that he had worked on, and many of his missions were still classified. Needless to say, I consider him to be one of the most intelligent people that I know both technically and scientifically.

As I listened to Papa recount his medical history with the doctors in the Florida market I felt my blood pressure begin to rise. Here sat a man who had kept his own medical charts, tracked his own blood count, and knew more about his body than most of his peers. With his engineering background, he had turned his technical expertise to attempting to figure out the world of medicine.

I began to understand that my grandfather, who had successfully chartered space missions to other planets, could not seem to get his own medical care under control due to lack of education about the medical system.

My grandfather explained the medication that he was taking and the side effects that he had experienced. He has a difficult time understanding the prescriptions given and did not know if he was taking a generic, branded, or other combination of the prescription drug.

With my knowledge of the pharmaceutical world and brief background in medicine, I quickly referenced the package insert for the drug as an attempt at help. I will never forget Papa and I reading the description together as I pieced together what he had been prescribed. I explained in layman's terms why he was feeling the way that he was, and how to communicate with the physician in "medical speak." It was evident at that moment the relief that my grandfather felt, as I eased his concerns. I assured him that he was not crazy and the doctor was prescribing Prednisone (a common steroid) for a good reason.

We decided to continue our conversation over time, and I always looked forward to answering the "tough" medical questions from one of the smartest men that I knew, over the phone or internet when time permitted.

I began to realize then that if a man who designed many of the space systems that are still in use today was having a difficult time navigating through his physician's office and hospital, then there must be a critical problem with the medical world in its present form.

Shortly thereafter my maternal grandmother, who was in perfect health at the age of 90, went to see a physician at Kaiser in California. She was told ever so gently by the healthcare provider that she was too old for the treatment and may want to look elsewhere. This was spoken by a physician to a woman with the energy and health of a 50-year old. My grandmother's life had taken her to Africa on safari at 80, and she still is taking Tai-Chi daily as of this writing.

Again, her experiences at the clinics in California made me think, surely there is a problem with the world of medicine.

The final straw came when I left my job as a pharmaceutical representative to pursue a career as a medical cardiology equipment rep for another very prestigious Fortune 500 company. I was placed over the D.C., Maryland, and Delaware hospital and medical market, and began selling equipment to some of the most prestigious hospital institutions in America.

Upon initiation I had very little training. A part of my job was to go into some of the top cardiovascular hospitals in America to assist them with what equipment to buy to scan their patients for heart disease. I saw a lot of economics being weighed in these sales negotiations, and very little patient concerns for the use of the equipment moving forward.

The management of the medical market that I witnessed during that time, along with my prior experience, made me passionate to try to make a difference in the world of medicine.

As I watched myself sell through relationships, watched the behind-the-scenes politics of the medical world, and witnessed poor medical education before my eyes, I thought to myself, surely there is a problem with the medical world, and the patient needs an education in order to better understand the environment.

This book was written to assist with the education of the average patient. We are entering into an era that has witnessed numerous problems in the medical world, from the Vioxx debacle to a medical epidemic of extreme proportions, as the baby boomers enter into retirement.

The stories from my grandparents, along with other patients that I have witnessed with my own eyes, showed me that the average patient needs to understand how to protect his or herself in order to navigate through what can be a confusing world of medicine.

Although much has been written about this subject from the experts in the medical world, I found that the books and online information fell short when really educating the average person on the foreign environment of the medical system. When I began in the field as a medical representative the terminology alone brought me to tears just trying to pronounce the words correctly. I can only imagine how frustrating it must be for the average person who is just trying to understand a correct diagnosis or treatment.

After my experiences as a representative and consultant, I decided put together a brief book that describes what the average patient should know in order to feel confident and comfortable that they are knowledgeable when entering a medical practice or hospital.

I took the perspective of a traveler in writing this book (one of my passions) because to any outsider the world of medicine can be intimidating, with its large words, multiple specialties, and technical pharmaceutical names, much as it would be to take a trip to a foreign country that you were unfamiliar with.

I hope that this book serves its purpose of my decade-long dream for Med-Ed, the education of the world one patient at a time.

Enjoy your journey.

CHAPTER ONE:

Keith Ellis, Bootstraps
Like the elephant, we are unconscious of our own strength. When it comes to understanding the power we have to make a difference in our own lives, we might as well be asleep. If you want to make your dreams come true, wake up. Wake up to your own strength. Wake up to the role you play in your own destiny. Wake up to the power you have to choose what you think, do, and say.

Welcome to the World of Medicine:

Welcome aboard, traveling patient.

You have come to the right agency for your visit to the world of medicine.

I will be your guide as we learn together how best to protect yourself through your visit to this confusing foreign land.

Here is a brief background for your upcoming visit:

The world of medicine continues to change at a rapid pace. Consider these recent headlines:

Medical mistakes are a stunningly huge problem, says a report by the Institute of Medicine. (Kohn L, ed, Donaldson, ed. To err is human: Building a Safer Healthcare System., Washington D.C., National Academy Press, 1999) It quotes studies estimating that at least 44,000 and perhaps as many as 98,000 hospitalized Americans die every year from medical errors. Even using the lower estimate, more people die from medical mistakes each year than from highway accidents, breast cancer, or AIDS.

Moreover, while errors may be more easily detected in hospitals, they afflict every health care setting: day-surgery and outpatient clinics, retail pharmacies, nursing homes, as well as home care. Deaths from medication errors that take place both in and out of hospitals – more than 7,000 annually – exceed those from workplace injuries.

The possible consequences of prescribing the wrong drug are grave: Epileptic patients receiving the anti-fungal drug Lamisil by mistake could experience continuous seizures. Patients erroneously receiving the antiepileptic drug Lamictal might experience a serious rash, blood pressure changes, or other side effects.

Errors also have occurred in prescribing the arthritis drug Celebrex, the anticonvulsant Cerebyx, and the antidepressant Celexa. There have been well over 100 reports of confusion among the three drugs, none of which has resulted in serious harm to a patient.

In one case, a physician wrote a prescription for "Celexa 200 mg." Since the antidepressant drug is available in only 20 and 40 milligram doses, the doctor was called, and he corrected his prescription to the intended Celebrex 200 mg. In response to such reports, the co-marketers of Celebrex, G.D. Searle & Co., Chicago, Ill., and Pfizer Inc., New York, have undertaken an educational ad campaign to alert health professionals to the possible mix-ups.

Most recently, The Associated Press reported that drug errors injure more than 1.5 million people a year. The most incredible piece in the report by the AP was that on average they estimate that a hospitalized patient deals with at least one medication error per day (AP, Drug Errors Injure more than 1.5 Million a Year, July 20, 2006; reported by MSNBC).

Based on these statistics and others, it reasons that we are moving into an era where each patient needs to understand his or her own clinical history, and to have a brief understanding of the world of medicine in order to feel confident that he is receiving the best quality care. I often compare where we are today in the medical world with where we were in the 80's and 90's in the stock market. Prior to that time, few individuals really managed or understood the 401K fund or managed their own futures. Pensions were still available in most companies, and stocks were considered a great addition when managed by brokers and Wall Street analysts.

Then things changed…

The corporate world became more volatile, pensions were not as common, and corporate retirement programs offered less of a guarantee than they had in the past. Every man became responsible for his or her retirement, and the 401K plan became a mandate for many with the hope of attaining retirement.

Suddenly, each person was responsible for having a working knowledge of mutual funds, stocks, and the ability to discern what percentage of his paycheck to give back to their retirement fund. As history recounts, the dotcom craze was born, online day-traders became popular, and coffee table conversations could speak to the stock market conditions of the day.

In my opinion, this analogy offers similar trends toward the future of medicine. Headlines have shown us that there is little guarantee anymore of "quality education and care" in all hospitals and medical clinics. Drugs have been pulled from the market, stories abound of potential problems with pharmaceuticals, and medical malpractice costs continue to soar.

Additionally, doctors' time has become scarcer, patients are less patient, and physicians do not get the same respect that they once garnered from their patient populations. As we peer more deeply into the world of medicine, the average person might begin to realize that medicine is less of a science and more of an art than previously thought.

Today's patient needs a working knowledge of the medical world in order to be aware of his rights, medical needs, and to assist with determining how to navigate through the system with success.

CHAPTER TWO:

Hellen Keller
Security is mostly a superstition. It does not exist in nature, nor do the children of men as a whole experience it. Avoiding danger is no safer in the long run than outright exposure. Life is either a daring adventure, or nothing.

Upon Arrival into the World of Medicine: Visit to the Doctor's Office:

The first potential problem that a patient usually experiences upon arrival to the world of medicine can commonly be called "white-coat syndrome." Most patients have been taught from our American medical culture that whoever wears the white coat is considered to be the experienced professional or practitioner. This can make for a nerve-wracking experience when visiting someone who is considered an "expert" in a field that the patient has little knowledge or experience in.

After my experience in the world of medicine, I suggest to the average patient some advice before taking on the "white coat assumption" upon arrival.

First, let's be honest: a white coat on any individual just honestly means that the person on the other side of the room most probably went through a lot of school, extended training, and paid a ton of money to wear a long white robe with authority. Although this is impressive, physicians put their pants on just like everyone else in the morning hours, and are often very approachable despite their high-level knowledge.

After my experiences in the medical world, I thought it would be helpful to explain how most doctors get to wear a white coat. In my opinion it is important to understand the training behind the person or people offering you treatment in order to gain knowledge into their realm.

Here is a bit of history on what it usually takes to become an MD in today's world of medicine.

The average physician in practice today most probably began by graduating near the top of their high school or college class. He or she usually attended four years of college, three to four years of medical school, one to two years of residency, and maybe did some post-graduate work. If they continued on in study, this would mean one to three years post-fellowship, additional certified medical coursework, and possibly another graduate degree or post-doctoral degree. Needless to say, this is a very well-schooled bunch.

It is important to remember that despite high education levels, physicians are approachable, and are in practice to help you to learn more about your condition or need.

As a word of advice, many physicians and patients in the world of medicine do not have the best communication skills. The lack of communication due to differences in terminology and educational backgrounds can lead to difficulty when a patient is trying to understand his medical care.

Additionally, there can be a bit of high-level expertise which physicians attain along with wearing a white coat after that much schooling. At times this leads to a feeling of an "earned right" to speak with authority, show control, and to be the expert when diagnosing the patient. Many physicians do not mean to, but end up talking over a patient's head when explaining medical protocols and drug therapies.

The truth is, statistics have shown that the average physicians' knowledge of current medical practices can actually begin to become outdated within as little as 5 years after graduation, due to the advances in medicine today. This is why it is critical for the patient to understand his or her medicines, health background, and medical condition.

In my experience as a representative, it never ceased to amaze me that I could walk into a physician's office and actually know more about medical indication or medication than the doctor that I was there to see. By considering the focus and in-depth study that I had been trained to have on a recent medical advancement, it was not surprising that at times I was the more educated person about a certain drug of the two of us. This was not because the physician was not intelligent in his area of expertise; his time was just very limited. There is so much new information on the market, it is difficult to stay abreast of every new drug or procedure that moves into the marketplace, unless it is your only job.

Second, as I mentioned earlier, both physicians and patients can, at times, be poor communicators. Statistics show that the average office visit lasts somewhere between ten to fifteen minutes in an internal medicine office. Most of the time, this does not offer enough time to give a complete medical history from patient to doctor. The burden is on the patient to communicate his or her needs completely in order to make sure that they are met.

In the world of medicine, the physician should not have to be the one to draw additional information from the average patient. Especially within time constraints. It is up to the patient to come prepared, be aggressive, and above all to be honest, to get the treatment which they deserve. In other words, YOU, the patient, are responsible to make sure that the communication of your visit is quality, and you must be the one to direct it that way from start to finish.

Third, if the statistics in the beginning of this book are not enough, the medical world also, at times, can be flawed. Just like any other business, mistakes can be made; things can go wrong, human error can occur. Patient charts can be misplaced or incorrect, clinical studies often do not offer large enough patient populations to make them completely accurate, and pharmaceutical drugs hit the medical market so quickly that they can become outdated prior to their scientific follow-up trials.

Over the years doctors have developed certain protocols for each patient's care. Due to lack of time, these protocols can be rushed, inaccurate at times, and difficult to manage.

It is up to the patient in today's world of medicine to demand better care and make each physician be accountable for the protocols, communication, and treatment that they ultimately deserve.

In all fairness to doctors, physicians can have a difficult time giving quality care in today's medical world. The demands of the healthcare system and the strains of reimbursements have created a tough environment for wanting to do a "good job" knowing you have to complete it quickly and effectively.

New medical school classes of today are teaching courses to graduating doctors about having a good bedside manner with patients, communicating effectively with patients, and managing medicine in a rushed and changing world.

With the lack of communication and hurried office visits, getting to the symptom of the patients' problem accurately in a short timeframe can become very tough. Add in the misunderstanding of the average patient into the terminology and education of the medical world and it is a disaster waiting to happen.

As mentioned earlier, many patients are too embarrassed or too intimidated to be forthcoming with their true symptoms when entering an office or hospital. The story that I heard as a medical representative is a good depiction of the average doctor visit of an embarrassed patient.

Here is a regular check-up situation: Mr. Smith walks into see Dr. Jones complaining of a head cold. Dr. Jones does a complete check-up of Mr. Smith, writes a prescription, and is about to leave the room when he asks Mr. Smith if there is anything else that he would like to add to his visit. It is at that time that Mr. Smith mentions that he is also having a problem getting an erection and wants some Viagra.

Clearly, Mr. Smith was not forthcoming with his true medical condition or his needs. The head cold which he initially mentioned probably was not the true reason for his office visit, right?

While direct to consumer advertisements for overactive bladder, erectile dysfunction, and other "embarrassing" issues have made patients more aware and more willing to discuss some of these common issues, there still tends to be communicative failure from patient to physician at times in areas of a discrete nature. We still have a long way to go to create good communication between patient and doctor. Taking control of your care is critical to having success upon arrival and beyond.

CHAPTER THREE:

Brian Dyson, CEO of Coca Cola Enterprises from 1959-1994 Imagine life as a game in which you are juggling five balls in the air. You name them - work, family, health, friends, and spirit - and you're keeping all of these in the air. You will soon understand that work is a rubber ball. If you drop it, it will bounce back. But the other four balls - family, health, friends, and spirit are made of glass. If you drop one of these, they will be irrevocably scuffed, marked, nicked, damaged, or even shattered. They will never be the same. You must understand that and strive for balance in your life.

Preparation for your Trip into the World of Medicine:
What to Bring, Clothes, and Documentation:

The best way to control your medical care in preparing for your trip is to go into the world of medicine as if you are on a mission. Just as you would not travel to a foreign country without a passport and extensive information about your destination, you should not venture into a medical environment without doing a complete check-up on your place of arrival.

As a medical representative, I was constantly trained to know that the doctor's time was scarce and valuable. I had to get my point across quickly and passionately while maintaining professionalism at all times.

As reps, we were trained to have a prepared elevator speech for the physician for most sales calls. Additionally we would prepare a longer version of that speech, just in case we were able to garner more than 10 minutes of time.

I offer you the same advice.

For a quality patient visit, my suggestion is that the average patient goes in with a plan, and understands how to travel through the medical system with an unwavering strategy.

Here is what you should have prepared prior to your office visit arrival:

1) Know your records. Have complete records of everything that you have experienced medically over time. This includes vaccines, medication lists, personal indications, any cancers, heart disease, unusual allergies, etc.

Be overly specific.

It is better to be more detailed than to miss something important.

Have everything written down prior to your arrival. Document what your symptoms have been as well; dates, times, any medications you have taken, etc.

2) Know your genetics: With the human genome project completed, medicine is facing a new era. Genetics and genetic therapies are on the forefront of the next era of medical care.

In other words, if your Aunt Marge had heart disease, it is a precursor on how your body might react long-term to certain foods, conditions, and whether or not you are predisposed to certain problems in the future.

List all relatives on both sides, including siblings, cousins, and children. Family history is critical in knowing where to begin for the practitioner.

Your doctor will consider this a roadmap to who you are and where you have been, and where you are possibly going.

3) Know the basics: Understand your physician's office dynamic, know what your medications are, know the basics of the

medical world (this book should help), and bone up on your symptoms and condition prior to your arrival.

Knowledge indeed is power.

The more you can speak the physician's language and understand possible treatments, the more inclined he or she is to give you more information or more time to discuss your symptoms and options.

4) Be preventive in your lifestyle. If you smoke, try to quit. Eat healthy food, demand more from your body, exercise, and take care of your best asset.....YOU. The more you work at preventing against diseases and health issues, the better your chances to stay on top of your game.

Also, it is important to understand that physicians are taught in medical school to treat through the case study method.

In other words, if you are a smoker, you will be placed into the "smoker" category of patient which they learned about in medical school.

Be honest, but understand that your patient history will place you among others in that category of health because medicine is based on case study and precedent.

5) Communicate, communicate, and communicate. Go into your visit on a mission to succeed. You are there not to just get treated, but to be treated well.

Demand proper communication from your appointment.

DO NOT leave until you get what you came for and get your questions answered. It is up to you to control your relationship to insure that your needs are met.

Consider the time with your doctor like a dating relationship or a marriage.

In order to get what you want, you have to ask.

Ask thoughtfully, but forcefully. If you are not getting your needs met, move on and find another doctor who will better meet your needs.

6) Learn to wear your own white coat. If it helps, wear your own white coat into your visit. Although I am saying this somewhat in jest, it is important to mentally don a white coat when you begin your visit with a physician.

Do not let white coat syndrome intimidate your ability to ask questions firmly and confidently.

When in doubt, ask the physician what he would do if he were treating himself, a family member, or his or her spouse. Sometimes this can make a difference in the quality of care that you receive.

As a recap, prior to your travel into the medical world, know yourself and know your lifestyle. Be honest, thoughtful, and forceful about your needs.

If you do not tell the truth about your health, you are only cheating yourself and affecting the quality of care that you will receive down the road.

Again, compare a visit with your physician to a travel to a distant land. If you are dishonest about who you are, you will probably be found out eventually. It is painful to try to be someone that you are not. Plus it could ruin your trip!

It is YOUR visit; take control of the itinerary.

Second, bring an accurate and complete list of all medications-the duration which you have taken them, any side effects, and all medical records from all the physicians in your medical history.

Third, Aunt Marge's heart disease was VERY important. Keep a detailed medical history of family members and lifestyles.

Fourth, do not allow your time to be compromised or rushed. You paid for your time. You deserve to get the treatment which you deserve.

Fifth, the accountability is unfortunately yours. You are responsible for your care, your time, and your follow up. Take detailed notes and bring a friend if necessary.

As an aside, patient advocate representatives have become a trendy addition to medical care as medical visits have become more difficult to understand. An additional person or associate in your favor may prove helpful if your condition warrants it.

As the old adage goes, four ears are always better than two.

A patient advocate can be a nice option if you need another person to help you through the visit and assist you in answering your questions.

Hold your physicians accountable to do the right thing and give you the best treatment available. Doctors usually want to do a good job, and just like any other relationship, with time and accountability, they learn to get to know you and your needs better. You must learn to use your time in the world of medicine wisely. You only are allotted a short time for your visit, so make it productive, educational, and memorable.

CHAPTER FOUR:

Hannah
Life is good, without it we'd all be dead.

Communicating with your Physician During your Visit in the World of Medicine:

So you have your script and your medical history in preparation for your travel, now how do you get it right once you have actually begun the trip?

As a medical and pharmaceutical representative, part of my training was to "role-play" an elevator speech in order to prepare for time with the physician prior to arrival. I would practice for hours in front of hired actors at our corporate office, the mirror in my bathroom, or other medical representatives in order to "get it right" when we arrived at the actual physician's office.

This role-play and continuous practice insured that we would not be intimidated when a white coat whipped around the hallway of a high-volume medical clinic.

I offer you this piece of advice because the medical world and the terminology of medicine can be intimidating. The land of medicine can be a scary place to navigate through even with a bit of practice prior to arrival.

The Drug names and Latin terms which the medical world uses can be confusing, and can be terribly difficult to pronounce and interpret.

For example, I marketed a quinilone antibiotic in my first sales position which had a difficult name to pronounce. Even the medical personnel that I called on mispronounced the name due to its confusing spelling and odd verbiage.

To insure visitation success, you should get into practice for your visit.

Prior to arrival, get your prescription drugs on paper, practice taking control of your vocabulary, and role-play mentally and verbally more than once.

Pretend that you are learning the exotic language of Medicine for an exciting trip. The more that you understand the local tongue, the easier it will be to control your visit and communicate with the natives.

Remember, knowledge is power. You are in better control when you walk into the medical clinic or hospital when you understand the facility and the physicians.

If you get into the right mindset and prepare prior to your visit, most of the details will be covered when you walk through the front door.

Be knowledgeable that when visiting a doctor or hospital you may not feel 100% healthy mentally or physically. It can be wise to bring along a friend in order to be confident and prepared enough to voice your history without falter. If you were visiting a foreign country, it would be totally appropriate to bring a tour guide for certain days, itineraries, etc. A visit into medical world is no different.

Your life is important. Go in armed with your records, your information, and your pride.

Commit to perfection, and do not leave until you have gotten what you need.

Be knowledgeable that life and accidents do happen. With preparation, you can stay in the drivers' seat. Education and preparation are the most powerful tools that you have.

Since I have made each office visit sound like a battleground, it is important to mention that a personal relationship with the medical staff and physician can go a long way toward better treatment. Just as knowing one of the natives on a foreign trip can make a difference in understanding the culture of the country, so too can having an insider's perspective assist with the scary world of medical care.

After all of the tough talk of the prior paragraph, I want to reinforce the fact that doctors and medical personnel above all are people with good hearts, and generally have good intentions in their vocation. They want to be respected, enjoy their patients, and they are known to enjoy a good joke or conversation if time permits.

It is important to remember the human side of the medical practitioner. They are experts and intelligent clinicians, but they have bad days, things go wrong, staff members get feisty. Be informed upon walking in, respect the time and management of the office and staff, and be patient towards things which are out of the physician's control.

As an aside, the gatekeepers of the office (secretary, nurses, etc) can give you a better understanding of your doctor's personality, lifestyle, and demeanor. The office staff often sit behind a desk most of the day, love to talk when time permits, and enjoy passing the time with anyone who takes interest or will listen to their daily woes.

It is important to try to develop a schedule with your office or hospital and develop good relationships with the physician and staff. If you can get a personal referral from another physician or someone else in medicine, it can work to your advantage by having a warm introduction rather than walking in without any background or history.

Just like any other network, the more that you work to develop a relationship with the physician and staff, the easier your visits will become, and you may even begin to feel more like a friend or close relative upon arrival.

Some of the checklist items to bring for your visit are as follows:

1) A detailed health profile- Questions about your profile that should be written prior to arrival

2) X-rays, reports, and medical records (if necessary)

3) Records and/or charts from any prior visits with another physician (if necessary)

4) Insurance information

5) Prescription medicine information, over-the- counter drug information, vitamins, and any other items that might be helpful. Also bring any vaccine dates and herbal supplements information.

6) A tape recorder, a friend, or a legal pad to take detailed notes of the appointment

7) Any additional questions which you want answered upon your visit.

Prior to your departure, it is critical that you ask yourself some questions to insure that you get the most out of your trip. Not everyone wants the same thing out of their medical care. Some people are more interested than others in getting specific details out of their visit into the Medical world.

Sample questions to ask prior to departure may include:

1) What kind of physician am I seeking? Are you most comfortable with a sympathetic listener, an expert in his or her field, someone who wants you to be a partner in your care, or one who reassures you in making decisions for your care and recommends a course of action?

What you get out of your trip is entirely up to you and how you want to have your needs met. The important thing is getting the details resolved prior to your entrance into the land of medicine.

2) Ask yourself why you are going in for medical advice. Is there something causing you to go seek out medical care?

3) What are your goals for the travel or visit? How will you deem the trip to be a success?

Good communication? Good feedback? Nice bedside manner?
This is important to lay out prior to departure.
Don't forget your passport!! ☺

CHAPTER FIVE:

Lowri Williams
Your life lies before you like a path of driven snow, be careful how you tread it cause every step will show.

The Natives in a Foreign World: Medical Personnel 101: MDs, PA, and DO (s) and all of the Other Initials that you May Run Into

Who exactly are you going to come across as you travel into the world of medicine?

You may run into a number of different indigenous people with different letters behind their names.

Here is a guide to the natives and what all of those initials stand for.

Personnel-

You will see many different natives in the world of medicine today. In the average medical facility, there are numerous individuals who may care for you in just one visit. These people take on many different abbreviations as they come out of school.

Although sometimes confusing, just understanding what all the initials behind the names of the natives mean, can be helpful.

We will start with the most common of the natives and work our way down from there. Enjoy the education.

M.D.-A doctor of medicine who has completed an accredited medical school and usually has gone through four years or more of medical training

D.O.-A doctor of osteopathy who has completed accredited medical school training and usually looks at the "whole person" rather than just a symptom or diagnosis.

P.A.-A Physicians' assistant is someone who has completed a two to three year accredited medical program who is licensed to practice under the care of a physician. P.A.'s usually have the same ability to diagnose and treat medical conditions as the primary physicians, with just a bit less authority.

N.P.-A Nurse practitioner is someone who has completed a RN (registered nurse) degree, who has completed advanced training in the diagnosing and treatment of medical conditions.

PT-A Physician Therapist is someone who has completed a one to two-year program in advanced medical training who usually works with designed exercises and medical equipment to help patients regain muscular function and improve physical abilities.

OT-An Occupational Therapist is someone who has completed a one to two year program in advanced medical training in the evaluation of joint conditions and the impact that the disease has on daily conditions. OT's design and prescribe assistive devices for patients which can assist with joint and muscle function.

Hospitalist-A hospitalist is a hospital-based general physician who usually admits and treats patients.

Chiropractor-A Chiropractor is someone who has gone to school to manipulate and adjust specific parts of the body in order to diagnose and treat a patient based on the belief that the nervous system is the control system of body function.

Massage therapist-A massage therapist is someone who practices therapeutic massage.

Herbalists-an herb doctor, someone specializing in the treatment of patients using herbs.

Nurse midwife-A nurse midwife is someone trained in nursing midwifery.

R.N.-A registered nurse is someone who has completed a two to four year program in nursing.

Pharmacist- A pharmacist is someone who has been trained in the science of drug therapy and prescription filling.

It is important that as a patient you check the credentialing of the person who is issuing you treatment. Make sure that you are focused on who you are seeing and ask for a second opinion if necessary.

These are the most common cast of characters which you will run into in the medical world. In today's changing realm of medicine, specialties and new positions are being created constantly. If you do not understand the personnel, ask the physician or staff to clarify who each person is and what they do.

If you need further information about what someone does in the office, it is completely appropriate to ask the questions about what exactly they do and how they fit into the overall picture.

Again, your visit is only as good as you make it in the medical world. The more effort you put in, the better your outcomes usually will be.

CHAPTER SIX:

Aaron Antonovsky, Unraveling The Mystery of Health: How People Manage Stress and Stay Well.
We are coming to understand health not as the absence of disease, but rather as the process by which individuals maintain their sense of coherence (i.e. sense that life is comprehensible, manageable, and meaningful) and ability to function in the face of changes in themselves and their relationships with their environment.

Different Lands Abroad: So Many Hospital Types, so Little Time:

The medical office or clinic of today has taken on a different form all together than it did in the past. In the past, a patient usually met with a doctor, a nurse, or possibly another medical personality during their visit. Usually this took place in an institution, hospital, or office setting.

No longer is there a one-size-fits-all philosophy in treating patients. There are outpatient hospitals, in-patient facilities, and community hospitals with numerous sub-specialists, physicians, and groups that are changing "traditional medicine" as we know it.

Here is a how-to on navigating your way thorough the changing environment of the medicine world from an eagle's eye view.

The follow list encompasses the basics, from a top down and very limited perspective.

A few different locations where you may receive treatment are as follows:

1) University Hospitals - Large, not-for-profit institutions that are usually affiliated with a university and/or accredited medical school.

2) Community Hospitals - For profit and not-for- profit hospitals which are not affiliated with medical schools and serve local communities in the areas of general medical and surgical services.

3) Off-site clinics - Medical treatment centers which focus on a limited number of specialties or specialty.

4) Children's' Hospitals - Specialty hospitals which focus on pediatric patients; this includes some university-affiliated hospital systems.

5) Rehabilitation Hospitals - Specialty hospitals which provide services to patients experiencing disability with things such as head injury, neuromuscular disability, and stroke.

This list could go on, but these are the basic facilities which the average person could see and be treated in at the time of this writing.

CHAPTER SEVEN:

Mahatma Gandhi
My life is my message.

How one who Graduates into the Title "MD" is Inducted into the Medical World:

During your visit, you may be curious as to how one becomes a native in the world of medicine. What exactly happens from the time a person goes to high school or college and then becomes an M.D.? Here is a brief translation into the medical training or background of the doctor that you might come across during your visit.

Native background of the average trained M.D.:

Internship-the first year of training following medical school.

Residency-a period of three to seven years of specialized training post-medical school.

Board-certification-a doctor has passed some exams given by the American Board of Medical Specialties in one or more of twenty-four specialties.

Licensure-The legal permission to practice medicine in a particular area granted by state or territory after an initial exam in that specialty.

So how do you navigate where your physician came from in deciphering all of this schooling????

First, you can check to see if a physician is board certified by the ABMS (American Board of Medical Specialties) by calling ASMS at 866-Ask-ABMS (275-2267) or visiting www.abms.org.

I learned in a very short period of time in medical sales to navigate successfully through the thicket of medical personnel. You must be informed, be prepared. If you do some research on your physician beforehand on their training, background, etc., it will go a long way toward understanding who is actually treating you.

As a helpful hint, keep your eyes and ears open upon arrival into a doctor's office or medical facility. Many of them will have their accreditation or background on the walls by way of a diploma or certificate.

This will give you an understanding of where someone trained and by what authority they can assist you with the details of your health.

Other helpful ways to look up or do a check on a physician are through web sites such as locateadoc.com, www.quackwatch.org, ABMS.org, or the American Medical Association.

Family referrals, friends, and other physicians also can provide great background information.

A good way to get a strong reference for a doctor is to ask another doctor who he or she would take one of his relatives to for treatment. Most people would not put their loved one under the knife without trust in the person on the other side of the handle.

As we discussed earlier, the most important thing is that you feel comfortable with the history of your physician and have a good relationship with him throughout your visit and treatment. The first step is the background check. Once you have been through this part of the process, you should be set with who you are ultimately going to see and why they are the best fit for your needs.

Additionally, a physician will probably be flattered that you know something about him, as most of them are proud of their background and schooling. Most, too, have one-way patient conversations all day every day and seldom have time to talk about themselves. It is nice to be able to discuss something relevant to them once you have done your due diligence.

CHAPTER EIGHT:

Hubert Humphrey
Life's unfairness is not irrevocable; we can help balance the scales for others, if not always for ourselves.

Medical Specialties Translated:

Just as the natives in the medical world have many years of schooling which they have attended, there are a ton of different specialties in the world of medicine to choose from. The natives have thousands of different tribes for the physician in the world based on your treatment needs.

Here is a brief overview of the most common, and what they actually do (in layman's terms), about what kind of doctor you are actually visiting:

Allergy and Immunology

An allergist-immunologist evaluates, diagnoses, and manages disorders involving the immune system. This can include such conditions such as asthma, anaphylaxis, rhinitis, eczema, and adverse reactions to drugs, foods and insect stings, as well as immunodeficiency diseases, defects in host defense, and problems related to autoimmune disease, organ transplantation, or malignancies of the immune system. As our scope of the immune system continues to develop, the definition of this specialty is increasing.

Anesthesiology

An anesthesiologist is trained to provide pain relief and maintenance, or restoration of the patient's condition during and immediately following an operation, an obstetric, or during a diagnostic procedure. The anesthesiologist also is responsible for assessing the risk of the patient's condition prior to, during, and after surgery. Additionally, anesthesiologists provide medical management and consultation in pain management and critical care medicine.

Cardiology

A cardiologist is a physician who is certified to treat problems of the cardiovascular system. This includes the heart, arteries, and veins. They are usually responsible for administering tests that indicate how well a person's heart is working, such as a treadmill or stress tests, and perform procedures such as cardiac catheterization, angioplasty, and echocardiography.

Colon and Rectal Surgeon

A colon and rectal surgeon is trained to diagnose and treat various diseases of the intestinal tract, colon, rectum, anal canal, and perianal area medically and/or surgically. This specialist also deals with other organs and tissues (the liver, urinary and female reproductive system) and how it relates with primary intestinal disease.

Dermatologist

A dermatologist is trained to diagnose and treat pediatric and adult patients with benign and malignant abnormalities of the skin (this includes skin cancers, melanomas and moles), mouth, external genitalia, hair and nails, and sexually transmitted diseases

Emergency Medicine

An emergency medicine physician focuses on the immediate decision-making and action necessary to prevent death or any further disability in the pre-hospital setting. This physician is assisted by the emergency medical technicians in the emergency department.

Family Medicine

A family physician is concerned with the total health care of the individual and the family, and is trained to diagnose and treat a wide variety of conditions in patients of all ages. Most recently, emphasis is being placed on prevention and the primary care of entire families, utilizing consultations, community resources, and genetic history.

Internal Medicine

An internal medicine physician is a personal physician who provides long-term comprehensive care in both the office and the hospital. He or she manages both common and complex illness of adolescents, adults, and the elderly. Internists are trained in the diagnosis and treatment of many conditions including cancer, infections, and diseases affecting the heart, blood, kidneys, joints and digestive, respiratory, and vascular systems. They are also trained in the essentials of primary care internal medicine, which includes education in the areas of disease prevention, wellness, substance abuse, mental health, and effective treatment of common problems of the eyes, ears, skin, nervous system, and reproductive organs.

Medical Genetics

A medical geneticist is a physician educated in diagnosing and treating therapeutic procedures for patients with genetically-linked diseases. This specialist uses modern types of tests called cytogenic, radiologic, and biochemical to assist in specialized genetic counseling, implementing needed therapeutic interventions, and provides prevention through prenatal diagnosis

Neurological Surgery

A neurological surgeon provides the operative and non-operative management of disorders of the central, peripheral, and autonomic nervous systems including the brain, spinal cord, peripheral nerves, and muscles, as well as the blood vessels which relate to these structures.

Neurology

A neurologist is a physician who has trained in the diagnosis and treatment of nervous system disorders, which include diseases of the brain, spinal cord, nerves, and muscles. Neurologists perform examinations of the nerves of the head and neck; muscle strength and movement; balance, ambulation, and reflexes; and sensation, memory, speech, language, and other cognitive abilities.

Obstetrics and Gynecology

A OBGYN is a physician who has skills and professional capability in the medical and surgical care of the female reproductive system and associated disorders. An OB-GYN has expertise in the areas of fertility, pregnancy, sexually transmitted diseases, and internal medicine for women's related issues.

Ophthalmology

An ophthalmologist provides comprehensive eye and vision care including diagnosing, monitoring, and medically or surgically treating all ocular and visual disorders. These physicians prescribe vision services which include glasses and contact lenses.

Orthopedic Surgery

An orthopedic surgeon is a physician trained in the prevention, investigation, and restoration of the form and function of the extremities, limbs, spine, and associated structures by medical, surgical and physical means.

Otolaryngology

An otolaryngologist is a head and neck surgeon who provides comprehensive medical and surgical care for patients with diseases and disorders which affect the ears, nose, throat, and the respiratory and upper alimentary systems. This specialty also can include related structures of the head and neck.

Pathology

A pathologist is a physician who uses information gathered from the microscopic examination of tissue specimens, cells, and body fluids, and from clinical laboratory tests on body fluids and secretions, for the diagnosis, exclusion, and monitoring of different diseases. Pathologists focus on the process of disease, like AIDS and diabetes, and are adept at using the tools of laboratory medicine to troubleshoot and solve problems. Pathologists try to improve outcomes and efficiency in medicine through scientific means.

Pediatrics

A pediatrician is a physician who treats the physical, emotional, and social health of children from birth to young adulthood. This can encompass a broad spectrum of health services ranging from prevention to the diagnosis and treatment of acute and chronic disease.

Physical Medicine and Rehabilitation

This specialty defines a physician certified in physical medicine and rehabilitation. Another term often used for this specialty is a physiatrist. These physicians are responsible for diagnosing and treating patients with physical disabilities. They evaluate and treat patients in order to achieve restoration of physical, psychological, social, and vocational function through comprehensive rehabilitation and effective pain management.

Plastic Surgery

A plastic surgeon deals with the repair, reconstruction, or replacement of physical defects of form or function involving the skin, musculoskeletal system, head and facial structures, hand, extremities, breasts and trunk. A plastic surgeon uses aesthetic surgical principles not only to improve undesirable qualities of normal structures, but in all reconstructive procedures as well.

Preventive Medicine

A preventive medicine physician is someone who focuses on the health of individuals or groups of individuals in order to protect, promote and maintain health and well-being. This can include preventing disease, disability, or premature death. Popular specialties in this field of preventive medicine most often specialize in aerospace medicine, occupational medicine, public health, or general preventive medicine.

Psychiatry

Psychiatrists are physicians who specialize in the prevention, diagnosis, and treatment of mental illnesses. This specialty can include mental, addictive, and emotional disorders. Mental illnesses, including addictions, are treated along with bipolar disorders, depression, and eating disorders.

Pulmonology

A pulmonologist is an internal medicine physician who specializes in the treatment and diagnosis of conditions associated with the lungs, breathing, lung diseases and abnormalities.

Radiology

A radiologist is a physician who diagnoses and treats diseases utilizing radiologic imaging procedures in adults, children, and special populations. Physicians practicing in the field of radiology often specialize in diagnostic radiology, radiation oncology, or radiological physics. These different specialties can utilize such tools as x-ray, radionuclides, ultrasound, and electromagnetic radiation to diagnose and treat disease, especially malignant tumors.

Surgery

A surgeon is a physician who manages surgical conditions which can encompass almost any area of the body. The surgeon establishes the diagnosis and provides the preoperative, operative, and postoperative care to surgical patients, and is usually responsible for the complete treatment of the trauma victim or the critically ill surgical patient.

Thoracic Surgery

A thoracic surgeon provides the operative, postoperative, and critical care to patients with pathologic conditions in the chest. This includes the surgical care of coronary artery disease, cancers of the lung, esophagus, and chest wall, abnormalities of the trachea, great vessels, and heart valves, congenital anomalies, tumors of the mediastinum, and diseases of the diaphragm.

Urology

An urologist is a physician who deals with diseases of the male and female urinary tract and the male reproductive organs. This specialist has complete education of endoscopic, percutaneous, and open surgery of congenital and acquired conditions of the urinary and reproductive systems and their contiguous structures, and is skilled in treating all of the conditions related to these areas.

Of course, there are many other specialties and sub-specialties which you might run into in the world of medicine.

These descriptions should give you an initial guide and glossary of which you may run into as you begin your trip.

CHAPTER NINE:

B. C. Forbes
The be-all and end-all of life should not be to get rich, but to enrich the world.

How Doctors Think and Speak: The Translator's Guide to Communicating with Doctors in the World of Medicine:

Learning the language and psychology of the natives in the world of medicine is not easy; it takes time and patience to understand the vernacular of this different world.

SOAP:

The natives in the world of medicine truly do speak a different language than the average person. As a medical world visitor you must learn to understand the psychology of the medical mind in order to get the most out of your trip.

When training to speak and communicate with a physician, it is critical to understand the way in which medical school students learn to treat patients during their study prior to entrance into the office or hospital.

Doctors go through medical school and are taught to communicate with each other and assess the patient by using what is known as the S.O.A.P. method. SOAP is an acronym for Subjective, Objective, Assessment, and Plan.

This method is critical for proper communication with your physician, because in order to get along with your physician you need to understand how to speak the language by which they have learned to diagnose.

Recent statistics show that anywhere from 70 to 80 percent of health problems can be diagnosed via patient communication with their physician. The odds need to be stacked in your favor as much as possible upon entry into the medical world, and the understanding of the SOAP method can assist with those odds.

Just as you would not go to a foreign country and try to communicate with the natives without brushing up on your vocabulary and the terminology for a new place, you should understand the commitment that it takes to get a point across to your physician quickly and effectively.

Here is the SOAP acronym interpreted so that you can communicate your needs in the native tongue successfully upon arrival.

S - The "s" in SOAP stands for Subjective. These are subjective observations which the patient verbally expresses or which are mentioned by a significant other upon arrival. Pain, discomfort, dizziness, nausea, and the like are part of the subjective definition.

O - The "o" in SOAP is Objective. This is anything which the physician witnesses, sees, touches, or learns through the initial check. Vital signs such as breathing, skin color, pulse, and respiration are included in the O category.

A - The "a" in SOAP is Assessment. This is the diagnosis of the patient's condition based on the "o" information.

P - The "p" in SOAP is to plan. Planning can include lab tests, referral to a specialist, patient directions (i.e. bed rest, time off, diet changes, etc), and follow-up visit directions.

The physician writes notes and communicates in this manner with his staff and colleagues.

Additionally, this documentation will include date and time of the visit, vital sign feedback from the exam, lab report information, and plan for the future.

The natives in the world of medicine are also taught in school by the "case study method." This means that when a medical indication or symptom is initially taught, the method studied is the outcome of an individual patient and what the protocol was for treatment success on that specific person.

For example, if a medical student were to learn about the common cold, he or she would learn about a 43-year old woman who came in complaining of sniffles, a runny nose, Kleenex usage, etc.

From there, the student would learn about the treatment protocols which helped this woman (the patient) get better.

This method is important to understand because it can help you to know that when you travel into the world of medicine, you are being compared to the 43-year old woman case study example in the mind of your physician.

This could be analogous to an American traveling to the African country of Congo. The natives of Congo have only seen a select group of American travelers, and so you therefore become the "same" as the prior visitors until they inspect you more closely and see things differently.

This is where your research into your visit becomes critical for success.

In Congo, you would want to have the experience and understanding of the prior Americans who had visited, but you would also want your specific trip to be personal and specialized to meet your own needs.

It would be up to you to show the Congan natives why you are different and how you can change their paradigm in order to be treated more uniquely than the average American tourist.

There are many helpful websites for Medical world information:

As a caution, the list of educational and medical websites can be LONG and varied. The average patient may become confused as he or she tries to uncover the truth in the vast sea of medical sites on the internet.

There is a lot of misinformation out there on the internet, and without a medical degree it can be tough to decide which internet site to believe or trust.

Again, consider your internet study into the world of medicine like doing research about a vacation to a foreign land. Although each guidebook or web site may give you a different perspective on your place of arrival, there is little time to read them all prior to departure.

The important thing to develop while conducting your research is an idea of what you are looking for, focus on the end result, and prepare the itinerary for your trip thoughtfully.

The important and more accurate information on the internet is usually repeated in most of the guidebooks and websites which you reference, so try to get a common theme about the information which you need to make the best educated guesses possible.

Here are some of the sites which will be helpful prior to your trip departure:

www.nlm.nih.cov/medilineplus/
www.healthweb.org
www.healthanswers.com
www.cdc.gov
www.mayohealth.org
www.mayoclinic.com
www.medem.com
www.medterms.com
www.webmd.com
www.medscape.com

Additionally, many pharmaceutical and medical companies offer assistance in navigating your way through your trip. You can reference their web sites and search engines for a more in-depth view of the medical world.

CHAPTER TEN:

Thich Nhat Hanh
Our own life is the instrument with which we experiment with the truth.

How to Get your Point Across in the World of Medicine:
Effective Communication with the Natives:

As mentioned, it can be a daunting task to try to communicate in the medical world. From my experience, you will get better results during your visit if you present a detailed, well-organized account of your present symptoms and give a detailed, relevant past medical history. The best way to accomplish this is to start with your most important symptom or problem at the beginning of your visit and work backwards toward your goal.

If you were visiting Paris, France for the first time, you would most probably go to the Eiffel Tower and the Louvre first, right? A visit to a medical office should be the same. Begin with the most important symptom and end with the small stuff.

As I have mentioned, the natives or medical staff in the world of medicine have their own language, culture, and protocols. Understand that you are on their turf and in their world when you visit.

You will garner a fair amount of respect if you show up with the knowledge of how to communicate, and a desire to learn from them about their culture.

If you are not getting the care you deserve upon arrival, you always have the right to switch your itinerary, see someone else, or choose another physician to communicate with. The world of medicine is a vast place. You may get a different perspective on your visit depending on which physician you talk to. The important thing is to get a well-educated physician whom you feel comfortable getting to know.

The more open you are about what your visit needs to encompass, the more you will get out of your travels.

Some other basic communication guidelines for physician communication are as follows:

1. Listen
2. Speak in "I" statements when describing your indication
3. Ask for explanations in simple language
4. Ask questions to clarify for understanding
5. Repeat what you have understood
6. Write down your instructions.

Similar to times when you have traveled on a vacation or a business trip, you probably can remember places and people, but over time after the end of your trip, the details usually begin to fade. This is why cameras, camcorders, travelers' journals, etc., are carried by most tourists on vacation.

I have learned on my travels through the world of medicine to take detailed notes and keep records so that I can remember the experience. Your visit to the medical world should be no different.

You want to have a detailed documentation of your trip so that you can share it with others or use it for a reference many years down the road.

Consider each visit an opportunity to further your education into the medical world and its inhabitants.

CHAPTER ELEVEN:

Anthony J. D'Angelo, The College Blue Book
Promise yourself to live your life as a revolution and not just a process of evolution.

What to Wear, What to Bring, and Your Demeanor upon Arrival:

The world of medicine offers a range of conditions and environments.

As the marketplace continues to change the world of medicine is also evolving.

The best way to mentally prepare for the climate is to use history and any personal references as a guide for your visit.

Prior to arrival be prepared for the busyness of most clinics and offices. Most offices and hospitals are packed to capacity with patients. Many are not very organized or efficiently run by the modern business standards of today.

Wear comfortable clothing during your visit. If you are more comfortable in tennis shoes or sneakers, wear them. If a business jacket makes you feel more professional, bring it along. Remember, you are traveling on your terms. Dress for success and to feel your best under the circumstances.

The climate of most medical offices tends to be fairly sterile and very busy. Although you will not be on your home turf, you can be prepared when you walk in, and as a result should feel more comfortable to act and be yourself.

Pay attention to the details of the office. Are the magazines arranged neatly? Is the staff dressed professionally? Some doctors forget that they are living in a retail world. Many of them could use a good lesson in the understanding of presentation, but again, this is a case-by-case scenario.

Be friendly and polite, no matter how long the wait for the visit. Try to maintain your professionalism, but be in control.

The key to finding the best treatment in the physician's office is usually to schedule your appointment early in the day.

An early day appointment means that the staff is usually fresher, the doctor is not running behind, and time is more manageable. As the day progresses most clinics and hospitals get a bit more frantic, crises occur, and your time or appointment may be pushed back.

Again, as I have stressed before, be prepared. Unless it is an emergency, be ready to wait awhile. Be knowledgeable of the best times and days for the practice or hospital, and get the gatekeepers to let you know when the best time is to get a good check-up/appointment.

You can use your wait time as an opportunity to think about what you want to learn from the doctor today. List the importance of these details on paper and review your notes.

The climate of each facility is different. You are a visitor, but with your new knowledge you are now a prepared visitor ready to enjoy the climate of your visit.

A busy medical office can be a good sign, not necessarily a bad one. Usually good doctors are the ones with the most experience and the packed waiting rooms can be a good indicator of a quality MD. Do not let the intimidation of a crowded office concern you too much. It is your opportunity to use your research survival skills to get to know the office manager and staff, and leverage a relationship for this visit or future ones.

As a representative, I learned that the office staff could be critical to a good experience in the office.

The office employees can be extremely knowledgeable; have usually been around for quite awhile, and with some personal attention, they can get you many of the details that you need prior to seeing the physician.

Get to know about the doctor through the staff. They have the ability to give the right information to anyone who builds a relationship properly.

Other words of advice upon arrival are to be on time, bring a smile; nice gifts are always appreciated (remember, these people are stuck behind a desk waiting on sick people all day everyday). It could never hurt to try to make them feel better, too.

You can get a pretty good feel for a medical practitioner or hospital by taking the time to interview the staff. If you have to wait, use the time to your advantage to add more knowledge to your travels.

An important thing to address in the search for a good climate is to look up a doctor or doctors by medical specialty or name. This can give you more background information on the native physician and his or her history. You can look many different places to locate good physicians, as well as climates, in your geography.

Here are a few to start with:

1) American Medical Association, www.ama-assn.org
2) American Board of Medical Specialties, www.abms.org
3) American College of Surgeons, www.facs.org
4) American College of Physicians, www.acponline.org

Again, much like traveling, not all physicians will be listed on each site. Cross-referencing is critical. If you find one on more than one site, it is a good sign that you are headed in the right direction.

CHAPTER TWELVE:

Joe E. Lewis
You only live once - but if you work it right, once is enough.

Precautions for your Trip:

There can be potential communication pitfalls on your visit to Medical world:

The scenery may be nice, but be prepared for being asked to further test the waters during your visit.

Your doctor may suggest that you get a test or tests to further clarify something uncovered during the initial S.O.A.P. consultation.

Again, it is a good idea to be leery of accepting the white coat syndrome of the locals without some questioning.

Ask, clarify, and understand what is being suggested prior to committing to any test.

There are some questions which you should have answered in order to insure that the test is necessary and helpful.

Here are a few basic q's to assist in the communication with the natives upon introduction:

1) What does the test measure?
2) What are the results for, and when will I receive them?
3) How is the test performed? Is there any pain associated with it? How should I prepare for the test?
4) How accurate is the test? Is there a margin of error? Who will read the results?

5) How much does it cost? Is it covered by insurance?
6) Why do I need it?
7) What are we ultimately trying to determine?

You can look up lab accreditation on certain web sites if necessary. Make sure that the lab is separate from the medical clinic, and ask why the lab was chosen for your tests.

During my experience as a representative there were, unfortunately, a couple of laboratories which were either owned by or were controlled by part of a physician group or practice. Although this does not mean anything specific with regard to the care which you will receive, it is a good idea to have a third party lab for your tests.

Most labs are not a part of this physician-owned situation, but caution is critical in insuring that you have a safe journey as you are asked to "test" the waters at your appointment.

www.labtestsonline.org can help to decipher what tests you are actually getting, what chemistry is involved, and what do the results actually mean. It is a great site for lab test translation.

CHAPTER THIRTEEN:

Albert Einstein
There are only two ways to live your life. One is as though nothing is a miracle. The other is as though everything is a miracle.

Destination Surgeryville:

Surgery can be a frightening destination as you travel through Medical world. No one enjoys the visit to Surgeryville, even those who have it planned and have scheduled it prior to their arrival.

Just like any of our other stopping sites along the way, the important thing prior to your visit to Surgeryville is to research your surgery center, the physicians, and the environment.

You can consider this research a scavenger hunt toward survival on your trip into this part of Medical world. This is where the TV show "Survivor" meets the average tourist destination. A patient advocate or other medical tour guide at times may be able to offer some help. A family friend, medical representative, or patient advocate might be good to include on your journey through Surgeryville, just for safety sake.

Consider this part of your journey the most exciting and adventurous one. Surgeryville is where everything that you have used in the clinician's office becomes more critical and important.

Be sure to do research on your surgeon and anesthesiologist prior to departure. Use the referral of your internal medicine physician or family doctor as a starting point and go from there.

In calling on surgeons and anesthesiologists as a representative, I learned that it was not always the most charismatic physician which had the best patient outcomes. Many top quality surgeons have different personalities, or are quiet personality types. They choose to make a living in operating rooms by choice, so this should not be overly surprising. It takes a smart, special individual to do this job. Don't be too quick to judge a surgeon just by his personality. The important thing is to have a surgeon who is well qualified and has shown good results over a quality period of time.

Any surgeon you choose needs to be board certified. You can check surgeon credentials at www.facs.org, or call 800-621-4111 for the American College of Surgeons template and information.

Prior to departure, be sure to look at studies, medical journals, ask friends and other medical representatives who they would suggest for your procedure. If you have time in the internal medicine office, ask some of the representative in the office what their thoughts are on doctor so and so. You will be surprised at the knowledge of the average medical sales representative into the community of doctors on which they serve.

If you know any anesthesiologists, this specialty is usually are a wealth of information when going to Surgeryville. The anethesiologists are the physicians who get to watch, and assist in all different types of surgical procedures, and are responsible for witnessing many of the details in the surgical process.

You can additionally check your state database and medical web sites to track outcomes of procedures. Many hospitals list their outcomes online so that consumers can get information regarding the standards of their facility.

Questions to ask again become critical as you enter into the land of Surgeryville. And, as before, documentation, notes, and follow up are critical;

Here are a few to get you started on your trip:

1) What is the surgical procedure?
2) Who will be performing the surgical procedure? As an aside, sometimes residents or fellows are able to do the surgery under the supervision of the surgeon. I would rather have an experienced surgeon doing the cutting unless it is a quick outpatient procedure.
3) What are the outcomes of this type of procedure?

4) What is the alternative if I opt against surgery?
5) What are the risks/potential side effects?
6) How many times/how long has the surgeon been performing this procedure?
7) How many of these surgeries does this hospital do a year? Do you have a record of the outcomes, or a grade for your facility?
8) How long will I be in the hospital? What is the downtime?
9) Do I need a friend or relative to help me drive, get home, etc?
10) What is the average recovery time?
11) What is the most common problem with this procedure? How does your hospital fare in comparison to local and national trends?
12) Do I need to fast? Is there pain medication available pre- and post-op?
13) What does the scar or long-term outcome look like?
14) Will I need any follow-up work, physical therapy, additional procedure, etc?
15) What is the cost? Is part of the procedure covered usually by insurance?
16) What kind of equipment are you using in the procedure? As an aside, as an equipment representative we sometimes tested out new equipment on the patient in the OR. You would rather have equipment being used which has been used many times prior. No need for one of the natives to be learning new procedures on your watch.
17) Do you have any pictures, videos, continuing education information that I can review prior to making a decision?
18) Who can I contact for follow-up visits, complications, and results?
19) How invasive is this procedure?
20) What is the pre-op and post-op protocol?

The land of Surgeryville can, at times, be a bit on the scary side. Going into a dark room, where the native locals are dressed differently and carry sharp objects, can get a bit intimidating for anyone.

Try not to get distracted, and do your homework prior to traveling on this mission.

Ask around, use the internet for study, and hold those in charge accountable until you get the answers that you deserve.

Additionally you may want to do things such as mark your own body (or the part to be the focus of the procedure), make sure that you meet the staff involved, and get everything on paper prior to procedure day.

The news has reported more than a few horror stories of patients who have received the wrong amputation, incorrect procedures, etc. Although these stories are few and far between, it never hurts to be safe rather than sorry.

If necessary, sit down with the surgeon and have him or her draw out the procedure for you to sign and review so that all parties are on the same page. Additional resources for research include the JCAHO website. This is the Joint Commission Quality Check information on how the hospital meets safety standards and performance measures. The site is www.jcaho.org, and is a good reference.

Additional things to know about Surgeryville are the cast of characters, the set up, and the knowledge behind the people who exist in those odd clothes commonly known as "scrubs".

Here is a brief rundown of what to expect when entering Surgeryville based on my representative experiences:

The surgery prep area is where a person is prepared for surgery. You will usually be asked to take off your clothes and put on a hospital gown prior to going in. You may meet with the anesthesiologist or surgeon at this time. Medications may be given before the operation. A person is brought to the operating room by wheelchair or on a portable bed with wheels, and then moved onto the operating table.

The operating table is at the center of the room surrounded by equipment tables. The anesthesia and monitoring equipment are kept at the head of the bed. This is where the anesthesiologist sits and monitors the person during surgery.

Assisting with the procedure in Surgeryville you will usually find at least one nurse and one technical assistant.

Caps, masks, and rubber gloves may be worn in the surgery environment. The room is sterile and is kept under strict standards for infection control.

In some types of surgery, the person is put totally to sleep with general anesthesia. In other cases, the person may be awake but a local anesthesia or regional anesthesia is used to prevent feeling pain. The surgeon comes in and the final preparation begins. The area of the surgery is often cleaned with an antibacterial chemical to insure cleanliness. A sterile drape is then put around the area of the surgery. Many people are already asleep or heavily sedated at this point.

The surgery is performed and can last anywhere from minutes for a minor procedure to hours for a more invasive operation such as open heart surgery.

After the surgery, the medication is turned off and the person slowly wakes up and is moved to the recovery room.

Surgeryville is not the ideal place to travel, but the natives are friendly and education will give you comfort if you happen to tour through this area of your trip. Be sure you document your travels to this land, it is important that you have all of the information just in case.

CHAPTER FOURTEEN:

Judy Newman
Hindsight is the "lottery" of life.

Choosing a Hospital: Navigating your Ideal Destination in the World of Medicine:

As we discussed earlier, there are many different types of hospitals to choose from when visiting the world of medicine. How then do you make an educated choice, about which hospital will best meet your needs?

The first intelligent place to start in determining which hospital might be best is to go to the Joint Commission on Accreditation of Healthcare Organizations (JHACO).

As a medical representative, I learned quickly to determine where each of my hospitals stood in the marketplace with JHACO.

JHACO grades hospitals on outcomes, quality care, and offers accreditation in report card form for hospitals and medical institutions.

Word of mouth from friends, relatives, or medical health workers, as I have mentioned, can also be extremely helpful in separating out the bad from the good.

Here is a checklist of some critical standards that any hospital should have prior to your arrival;

1) Is the hospital accredited by the Joint Commission on Accreditation of Healthcare Organizations?
2) Is it rated highly by State or consumer or other groups?
3) Does your doctor have privileges at that hospital; if he or she does not, you will need to be open to treatment by someone else.
4) Is the hospital included/covered by your health plan?
5) Does the hospital have good experience with your condition, and good outcomes?
6) Does the hospital have quality control procedures, and does it work to improve outcomes on a regular basis?

As mentioned, JCAHO accreditation should be your first step in choosing your destination. Think of JCAHO as the United Nations of your trip. They are the ambassadors of quality care and quality checks and balances. Hospitals can choose to be surveyed by the Joint Commission on Accreditation of Healthcare Organizations (JCAHO) to make sure they meet certain quality standards on a regular basis. The standards address the quality of staff and equipment, and the hospital's success in treating and curing patients with regard to outcomes analysis. If a hospital meets these standards it then becomes accredited. The review process is usually every three years.

If you go to www.jcaho.org you can check the hospital's performance report for it's accreditation status.

The hospital usually is graded on:

*Accreditation status (six levels - from the lowest, "Not Accredited," to the highest, "Accredited with Commendation").
* Date of the survey.
* Evaluation of the key areas reviewed during the survey.
* Results of any follow-up activity.
* Areas needing improvement.
* Comparison with national results.

Additionally, you can look for hospital report cards developed by states and consumer groups.

Ask your doctor or the hospital for information on:

* How often the procedure is done there.
* How often the doctor does the procedure.
* The patient outcomes (how well the patients do).

Choosing an Emergency Room (ER):

If choice of a hospital is confusing, how then does one choose which ER is the right one to visit? ER research should be done hopefully long before you have to visit the natives in this part of the world of medicine. Remaining in control and having knowledge to navigate through your trip is critical in this venture.

Get information from your doctor; ask where he or she would be treated if he had an emergency themselves.

Many of the checklist items (JCAHO accreditation, does your physician have privileges there, is your health plan accepted) are still important in choosing an ER destination of choice.

As a medical representative I learned about the rating of ER's (yes, they do get rated).

Also, I learned that a trauma center is completely different from an ER.

A trauma center designation sets strict requirements for staffing, specialist availability, response times, training, quality improvement, and community education. Additionally, trauma centers have organized trauma teams which respond promptly to trauma alerts, a surgeon who serves as trauma director and provides oversight to the hospital's trauma program, trauma nurse coordinators, and committees which provide quality improvement and direction for the hospital's trauma program.

The emergency departments of hospitals which are not designated trauma centers may be staffed by an emergency physician day and night, but do not have organized trauma teams ready to respond to a trauma call or access to the immediate, high level of surgical care available at a designated trauma center.

This is the brief difference between the levels of trauma centers that you will find. It is important to see which of these exist in your specific geography and have a game plan in store in the event that you need level 1, 2, or 3 treatment.

A Level I trauma center has the highest level of capabilities available.

Generally, these trauma centers are attached to medical schools or have on-site residency programs because of the in-house requirements. Requirements for a Level I trauma center include:

- In-house emergency medicine, general surgery, anesthesia on a continual basis;
- Other specialists additionally must be on-call and be able to respond within a short time to a trauma alert;
- The specialties that must be represented are usually as follows:
    - Cardiac surgery
    - Hand surgery
    - Neurologic surgery
    - Obstetrics/gynecologic surgery
    - Pediatric surgery
    - Ophthalmic surgery
    - Reconstructive surgery
    - Orthopedic surgery
    - Plastic surgery
    - Surgical critical care medicine
    - Radiology
    - Thoracic surgery

Trauma 1 centers are responsible for the additional resources to contend with if they are to keep their rating. These, as I learned, can include:

- Extensive equipment requirements;
- Specific clinical qualifications and trauma-specific continuing medical education requirements for physicians and other medical staff;
- Operating Room availability 24/7, and in-house OR staff;
- Specific quality improvement monitoring of trauma patient care, and continual monitoring of trauma care protocols and policies;

- Participation in injury prevention activities within the community; and
- Research requirements.

<u>A Level II trauma center has extensive capabilities to meet the needs of most trauma patients.</u>

I learned as a representative, the major difference between Level I and Level II facilities is that the major surgical specialties are required to be on-call in Level II centers, but with the clear commitment to be in the Emergency Department when the patient arrives.

Requirements for a Level II trauma center include:

- In-house emergency medicine and anesthesia capability at all times;
- Other specialists must be on-call and must respond within a short time to a trauma alert as follows:
  - General surgery
  - Neurologic surgery
  - Obstetrics/gynecologic surgery
  - Ophthalmic surgery
  - Reconstructive surgery
  - Orthopedic surgery
  - Plastic surgery
  - Radiology
  - Thoracic surgery

As with a Level I center, the additional requirements which must be kept up in order to maintain Level II status are:

- Extensive equipment requirements;
- Specific clinical qualifications and trauma-specific continuing medical education requirements for physicians and other medical staff;
- Operating Room availability 24/7 and OR staff on-call and available 24/7;
- Specific quality improvement monitoring of trauma patient care, and continual monitoring of trauma care protocols and policies; and

- Participation in injury prevention activities within the community.

A Level III trauma center is a hospital which is committed to caring for the trauma patient, and provides prompt assessment, resuscitation, emergency operations, stabilization, and possible transfer to a facility which can provide definitive trauma care.

Although the specialist and equipment requirements are not as strict for Level III trauma centers, these hospitals must provide prompt general surgical and trauma team response to trauma alerts. The care of the trauma patient is monitored by strict quality improvement. Level III trauma centers must meet the minimum requirements listed below, some specialists must be available as well, but it is mandated by the different hospitals rather than an outside governing body.

Requirements for a Level III trauma center include:

- In-house emergency medicine and anesthesia capability at all times;
- Other specialists must be on-call and must respond within a short time to a trauma alert:
    - General surgery
    - Radiology
    - Extensive equipment requirements
    - Board certification in the physician's specialty is required and a trauma-specific continuing medical education requirement is desired.
    - Operating Room availability 24/7 and OR staff on-call and available 24/7
    - Specific quality improvement monitoring of trauma patient care and continual monitoring of trauma care protocols and policies
    - Participation in injury prevention activities within the community

Additionally, here are some helpful checkpoints prior to venturing into this sector of your trip:

1) Find out if the physicians in the ER have been trained in emergency medicine. This is not always a prerequisite for working in the emergency department, but can be important.

2) Find out the background of the nurses. You can go to www.nursingworld.org to see if the hospital had earned "top status" from the American Nurses Credentialing Center.

3) Find out if the nurses have been trained and certified in ACLS (Advanced Cardiac Life Support), CEN (Certified Emergency Nurse) or PALS (Pediatric Advanced Life Support). The facility should be able to tell you this.

4) Find out the patient volume and outcome rates.

5) Ask about staff experience and specialty.

6) Ask about patient volume with regard to trauma, and amount of patients seen with your condition.

7) Ask about hospital support services (MRI machines, diagnostic imaging equipment, therapy, surgical procedures, CT scanners, and round-the-clock diagnostic tests).

8) If the facility is not a Level I, ask about the proximity to a Level I Trauma Center, or transportation possibilities.

If you have children, you may want a few more things answered prior to choosing a hospital to insure that your young ones are getting the level of care which they deserve.

You might look for:

- Close proximity to your home
- Emergency Medicine physicians also trained in treating children
- Nurses trained and experienced in treating children
- Separate, child-friendly waiting area
- Separate treatment area for children
- Equipment and supplies designed especially for children
- A hospital where your pediatrician is on staff so that your child's care won't be turned over to an unfamiliar physician

- Pediatric unit and/or pediatric intensive care unit within the hospital
- Pediatric consultants available in the emergency department

Additional websites for education for your trip to the ER include:
www.aha.org (American hospital association)
www.hospitalcompare.hhs.gov
www.leapfrog.org
www.pressganey.org
www.healthgrades.com

CHAPTER FIFTEEN:

Julia Child
Life itself is the proper binge.

Understanding Drugs:

The best way to think about pharmaceutical drugs in the world of medicine is to compare them to the souvenirs which you might receive from your travels abroad to a foreign land.

Interpretation of the Package Insert:

I learned as a medical representative that one of the ways to "win" in the world of medicine is to understand the scientific language of the package insert.

The package insert is the piece of paper which is attached to the drug, or is inside of the sample pharmaceutical, which explains all of the science behind the product. The problem is that the education level required to interpret this piece of paper is beyond most average individuals. Even as a representative I knew that I could have the upper hand when walking into the world of medicine by "quoting the p.i." Most physicians do not have the time to read all of the ins and outs of the product information and it gets confusing trying to keep all of the drug package inserts in order.

It is not necessary to be a bio-chemical engineer in order to decipher the language of the package insert, but one must at least understand the basics in order to read between the lines about where a potential problem might lie in taking a medication.

Here is the kindergarten version of what the parts of the package insert, or "p.i.," actually says and does. Consider this your alphabet whenever you have received a prescription from your physician.

As a word of caution, if you read the p.i. of any product and take everything to heart with regard to potential side effects and outcomes, you probably would not even ever take an aspirin for a headache.

Just a reminder, everything on the market (including Advil or Tylenol) comes with a risk. This guide will hopefully make the terminology a bit more manageable so that you do not misunderstand the risks, and can manage the rewards.

Here are the pieces of the puzzle broken down more easily.

The critical things to remember when receiving a prescription are:

1) You do not have to take the medication (it is up to you to weigh the benefits and risks accordingly with education);

2) Side effects are very important. For example, why would you want to trade a headache for a more serious side effect of a drug or treatment?

3) Drug-to-drug interactions can be very important to pay attention to. Hopefully you have been honest with your physician about what you are taking, but drugs do not always marry very well when combined;

4) Future effects and clinical trials can offer a ton of education as to the experience of the drug among patient populations and in the market.

The pieces of the package insert alphabet deciphered in layman's terminology:

Although every package insert can be different, these are the basic pieces that you will come across when opening up a drug for ingestion or administration.

The first thing that will be listed on any package insert is the branded drug name, the scientific name, and the description of the product (i.e. whether it is a capsule, liquid, tablet, etc).

1) Description - The description section of the package insert is where you will find the scientific background on the drug, along with the molecular chemical make-up and ingredient make-up. This information is usually only helpful if you were a biochemistry major or want to "Wow" your friends in trivia by knowing the scientific name for Viagra.

2) Clinical Pharmacology/Pharmacokinetics - This area of the P.I. describes how the drug breaks down and metabolizes once entering the body. This section again will put the average person to sleep, but is important in knowing how the distribution, chemical reaction, and metabolism of the drug takes place.

Pharmacokinetics refers to the study of what the body does with the drug, and what the drug does to the body. This area can be important with regard to potential for drug-to-drug interactions, metabolism with special populations (i.e. elderly, children, along with certain races or gender).

The half-life of the product describes how much time it takes for the amount of the drug in the body to be reduced by half strength. This can be important, because certain drugs can leave your system within 2 hours, others 2 days, and others 2 months. If you have a drug-to-drug interaction or side effect, this can be the difference between a short and long term effect and reaction to the medication. Additionally, this can affect the efficacy of the product. We would pay close attention to this section of the package insert as medical representatives, because it is important for the doctor and patient to know how long the actual drug will be in the body. As a word of caution, a drug may have a short half-life, but a long duration of effect. For example, the product Botox has a half-life of 48 hours, but the effects of the injection can last up to 4 months.

Pharmacodynamics: This refers to what the effect of the drug actually is. In other words, this is the study of the biological and physiologic effects of the drug and the mechanism of action. How does the drug actually work, improve performance, cure disease, etc.

3) Clinical Studies - This is the section of the package insert that is devoted to trials which were submitted for approval of the product to the FDA.

4) Indications - This refers to what populations and what indication the product or drug is actually developed for. This section can be important, because many products are being written off-label or off of the regular indication much of the time. This can happen for a couple of reasons; one, the doctor has developed his or her own protocol for a certain medication, or two, the drug company never sought additional indications of the product. This is not uncommon, and happens frequently as a drug increases in popularity. Some physicians stick strictly to the package insert, but most develop their own ideal of how to use common medications safely.

5) Contraindications This section is the part of the p.i. which discusses any patient types who should not be given the drug due to ingredient make-up, patient types, or clinical trial information.

6) Warnings - This section discusses any specific precautions with patient types or certain populations based on clinical trials or post-marketing surveillance information.

Black Box warning: This is the strongest FDA warning that the package insert can receive. It alerts the doctors and the public to serious health risks or problems which have been associated with the medication.

A drug only receives a black box warning if enough people have had negative, serious effects. This is the most serious warning to be given to any product or drug.

7) Precautions - Caution to the patient population, special population types, ingredient make-up, or any side effect noticed during the clinical trials which might be relevant prior to prescribing the medication.

8) Drug interactions - This section describes whether or not the drug has any reported drug-to-drug interactions which were noticed upon invention or during the clinical trials.

Information for patients: This section has specific instructions for patients taking the drug. This includes pregnancy, medical conditions, special populations (gender, race, age, etc.).

9) Pregnancy - This refers to the risk category that the product received during its initial trial phase. The categories of a drug for pregnancy can be five different options:

Pregnancy Category A: Adequate and well-controlled studies have failed to demonstrate a risk to the fetus in the first trimester of pregnancy (and there is no evidence of risk in later trimesters).

Pregnancy Category B: Animal reproduction studies have failed to demonstrate a risk to the fetus and there are no adequate and well-controlled studies in pregnant women OR Animal studies which have shown an adverse effect, but adequate and well-controlled studies in pregnant women have failed to demonstrate a risk to the fetus in any trimester.

Pregnancy Category C: Animal reproduction studies have shown an adverse effect on the fetus and there are no adequate and well-controlled studies in humans, but potential benefits may warrant use of the drug in pregnant women despite potential risks.

Pregnancy Category D: There is positive evidence of human fetal risk based on adverse reaction data from investigational or marketing experience or studies in humans, but potential benefits may warrant use of the drug in pregnant women despite potential risks.

Pregnancy Category X: Studies in animals or humans have demonstrated foetal abnormalities and/or there is positive evidence of human fetal risk based on adverse reaction data from investigational or marketing experience, and the risks involved in use of the drug in pregnant women clearly outweigh potential benefits.

10) Additional Miscellaneous categories/Information for patients:

Carcinogenesis, mutagenesis, impairment of fertility - This refers to the long-term risks of taking the drug.

Pediatrics - Whether the drug has been tested on children.

Geriatrics - Whether the drug can be used to treat elderly individuals.

Nursing mothers - Risk with regard to drug administration and nursing.

11) Post marketing - Any post-marketing data or trials after the product was released to the market.

12) Dosage and administration - The recommended dose of the product; specific dose adjustments, individualized doses, how to take the drug, dosage level, and administration directions.

13) How supplied - Dosage measurements, the color of the drug, the description of packaging.

14) Storage - How to store the drug and how long it is "good for" in storage. Also, this can have the dosage measurements, the color of the drug, and the description of packaging.

15) Side effects/Adverse reactions - The side effect profile of the drug from the initial clinical trials. This can include a chart of percentages from the number of patients who experienced side effects during the trial phase. As a word of caution, every side effect which any patient experienced during a clinical trial has to be reported in the package insert. This can be misleading and could show, for example, that a common cold treatment could cause a heart attack, when in fact the treatment which the patient received for his or her cold had nothing to do with the heart attack that they would have experienced with or without the drug treatment.

16) Drug handling - Container rules, shipment rules, storage rules, tamper proof rules. This also can refer to the use of the remaining drug if not used up during the initial treatment, length of storage, refrigeration, and procedure for proper administration.

17) Manufacturer and distributor info - What company makes and manufactures the drug, what companies create and market the product. This can be one or many companies due to joint marketing procedures, co-development, and mergers and acquisitions.

Some great website references for drugs and clinical trial information are:

www.clinicaltrials.gov
www.fda.gov
www.centerwatch.com
www.drugs.com
www.pdr.net
www.safemedication.com
www.webmd.com
www.usp.org
www.ismp.org
www.medicineplus.gov

Additionally, when receiving a prescription, you will see some Latin shorthand which describes how a pharmacist is to dispense the drug and how many times a day you may take it.

Without understanding Latin shorthand, which to my knowledge is not taught in any school, this can become very confusing to the average consumer. Here is your translation guide to decipher what exactly those Latin-looking abbreviations mean:

Ad. Lib. = As much as wanted; freely
AC = Before Meals
BID = Twice a Day
BP = Blood pressure
CC = Cubic Centimeters
CUM = With
DSP = Amount to Dispense
EN = Enema
G = Gram or grams
HN = Tonight
HS = At Bedtime
L = Liters
Oz = Ounce
PA = By the anus
PC = After meals
PM = As needed
PO = Through the mouth
PRN = According to circumstances, as needed
QD = Take Daily
QH = Every Hour
QID = Four times daily
SIG = Write on the Label
SINE = Without
SOS = May be repeated once
Stat = Start immediately
TAB = Tablet
TID = Three times per day
UNG = Ointment
2X = refill two times

Hopefully you are feeling a bit more fluent for the pharmaceutical souvenir which you may receive.

Let's keep traveling!!!

CHAPTER SIXTEEN:

Pericles
What you leave behind is not what is engraved in stone monuments, but what is woven into the lives of others.

Where do these "Souvenirs" come from, and why are they so Expensive?

Studies have shown that today 91 percent of seniors and 61 percent of non-elderly adults have to rely on some sort of a prescription medicine on a regular basis. Although still only a modest part of total health care spending in the U.S. (11%), with so many people relying on prescriptions the cost implications loom large for the American public, health insurers, and government payers. Additionally, the drug industry's profit margins have raised considerable attention to the situation at hand.

As we have discussed, the world of medicine can be a scary place. Statistics show that the drug marketing industry has gone from 40,000 to 90,000 sales representatives in about 7 years according to the company Verispan based in Yardley, PA. This means that there is a lot of information and a lot of people out there in this market. A lot of reps are doing a lot of selling to physicians who many times are too busy to read every line of every drug which they prescribe.

Bringing a drug to market is an expensive process which can vary in cost depending on who you speak to. Drug companies go through a fair amount of research and development at some point or another during the process, and usually are responsible for paying for some part or the whole of the clinical trial process.

Usually, clinical research is initially done in a laboratory or on a research lab basis, and then a drug or a biotech company will step in and play its hand in the process.

The most important thing for the average patient to understand is the background of clinical trials, and what standards have to be met prior to a drug's approval. Drugs are not just placed into a human mouth and given the okay from invention. It can be a long and difficult process, and depending on who you ask, the cost for a new pharmaceutical can range anywhere from 2 billion to 800 billion dollars. The figure itself has been controversial due to the drug company scrutiny as of late, but in all fairness, not all drugs which are put through the process are approved, and many are disposed of before they can ever see the light of day.

So let's go through the clinical trial process from start to finish and get a background on what a clinical trial of a regular drug might look like….

The development of the R&D process consists of two separate categories: preclinical and clinical trials.

In the initial preclinical phase, drugs are tested on animals or in Petri dishes in the laboratory. This can be done on cultures from plant, animal, or cell cultures. Very few drugs even leave the preclinical phase and move on into the clinical stage of the process.

Statistics by pharmaceutical companies are that one in five thousand candidate drugs even makes it to market. If the drug passes through the pre-clinical phase of the process, it is then passed on to the clinical stage where it is regulated by the FDA. A pharmaceutical or biotech company must then prove that the drug is safe and effective under certain FDA guidelines. Clinical trials usually are divided into three separate phases:

1) Phase I is where the drug is given to a small sample of volunteers in order to determine safe dosage levels and side effects.

If Phase I goes well, the drug is then moved into Phase II.

2) Phase II is where the drug is tested on a larger sample population who has the disease or medical condition which is being

tested. The drug is tested at different sample doses, and side effects and efficacy are tracked.

If Phase II is shown to be effective, the drug is then moved into Phase III.

3) Phase III is where the drug is evaluated on an even larger group of patients for safety and effectiveness. Not all drugs go through all three phases in this process, and not all phases are released to the public as a part of the information about the drug.

The latest trend in the clinical trial process is for profit and not-for-profit companies to set up trials rather than going just through medical schools and university hospitals.

After Phase III some drugs go into Phase IV post-marketing studies. These are to track how the drug does after it has already been released to the public. This serves to follow up for side effects or other expansions of the drug's indication.

As I learned as a medical representative, many of these "trials" need to be read very carefully. It is critical that the average patient does not just take a trial at face value, but instead looks at the entire patient population studied.

Drugs which only went through the Phase I and II parts of the study usually had very small populations, therefore the side effects and efficacy parameters of the drug could be skewed due to the small amount of people actually being studied. As mentioned, much of the time the clinical studies are paid for by the pharmaceutical or medical company. As a result, many studies are weighted in the direction of the clinical drug on trial.

CHAPTER SEVENTEEN:

Christopher Morley
There is only one success - to be able to spend your life in your own way.

Taking your Souvenirs Properly: Don't Drink the Water Until you Make Sure that it is Safe:

According to the groundbreaking 2003 medical report Death by Medicine, by Drs. Gary Null, Carolyn Dean, Martin Feldman, Debora Rasio, and Dorothy Smith, 783,936 people in the United States die every year from what are considered conventional medical mistakes. That's the equivalent of six jumbo jet crashes a day for an entire year.

Since Americans spend so much money on health care, they should be getting a high quality of care, right? Unfortunately, that's not the case. Of the 783,936 annual deaths due to conventional medical mistakes, about 106,000 are from prescription drugs, according to this study.

As frustrating as it can be to have to read a package insert prior to popping a pill, I have learned as a medical representative that “the devil is in the details" when making an informed decision about pharmaceuticals. Many of the medicines which are prescribed in the world of medicine today should be prescribed more carefully. The prescription drug industry should not be a one-size-fits-all production line for consumers.

Most physicians take this seriously. The important thing for anyone delving into the world of medicine is to use a cautious demeanor in order to protect himself.

As a representative I learned that there is power in knowledge and education.

This book can be your guide; use it!

There are some great questions which can be asked prior to taking your medication to insure that you are getting the desired drug and the desired effect that you deserve.

Here are some great ones to get you started:

1) What is the brand name and the generic name of this medicine?

2) How does the generic medication differ from the original formulation?

3) What are you prescribing this medication for?

4) Does this medicine require that I discontinue the use of any of the other medications or supplements which I am currently taking?

5) How do I take this medicine? How many times a day should I take it? Do I need to eat prior to taking it? Should I take it early in the morning or late at night?

6) When should the effect of the medication take place?

7) How long is the prescription for? What if I need a refill?

8) How can I tell if it is working?

9) Are there any foods, drinks, other medicines, physical activities, driving, etc, which I should avoid while on this medication?

10) What are the side effects that I may experience? What percent of patients experience these?

11) What if I miss a dose? Is there a way to "catch up?"

12) Do you have any additional reading material or brochures about the product?

CHAPTER EIGHTEEN:

N Smith

When you live in reaction, you give your power away; Then you get to experience what you gave your power to.

Clinical Trials Deciphered: So a Souvenir can be Tested by Ten People and be Considered a Worthy Drug????

Studies funded by pharmaceutical companies many times conclude that the best drug is the one made by the companies sponsoring the trials, according to a study in the American Journal of Psychiatry which was reported in the Washington Post (April 14, 2006). Psychiatrist John Davis and colleagues analyzed all publicly available studies funded by the pharmaceutical industry which examined the effectiveness of five new antipsychotic drugs. They found that nine in ten of the studies showed that the best drug was the one made by the company that funded the study. An example given was that Eli Lilly recently funded five studies which compared its antipsychotic drug Zyprexa with Janssen's drug Risperdal; all five studies say Zyprexa is a superior treatment for schizophrenia. When Janssen sponsored its own studies comparing the drugs, it found that Risperdal was superior in three out of four. On the basis of these contrasting findings in head-to-head trials, researchers concluded that whichever company sponsors the trial seemed to produce the better antipsychotic drugs.

Clinical trials have come under a ton of scrutiny as of late. Medical and drug companies have been accused in such trials as the Vioxx incident of forcing drugs "through the approval process" quickly, approving drugs with known side effect conditions, or allegations of trading money for clinical speediness in the drug approval process.

So, when getting a product from the world of medicine, even from a practitioner whom you trust, how do you interpret the clinical trial information to know whether or not that drug is a worthy one? There are a few basic questions which need to be answered prior to the popping of any pill which can many times be answered by interpreting the clinical trial information clearly.

Remember, just like a visit to a foreign land, each person has a different degree of risk aversion toward what they are comfortable with while they are there. The same is true with clinical trials. Many trials are done on patients who are willing to be first to market themselves, or are interested in an alternative treatment to what they could have had through the traditional medicines on the market.

There are benefits to being first to market, as well as deficits. The patients who are willing to try something first may have access to the newer version of a medication, or a next generation of a drug. The deficit is, as we discussed, the product has not yet been tested on all metabolisms and all types of individuals, and as a result, there may be things yet unforeseen about the drug's reactions, side effects, or long-term effects.

I am not going to offer advice through this guidebook on what to do with regard to the trials which are on the market. This, instead, should serve as a background so that you can choose your comfort level on what medications you want to take.

The sites, www.clinicaltrials.gov and www.centerwatch.com both offer not only backgrounds on clinical trials, but also can give you information if you would like to be a participant in an upcoming study.

So how does the average visitor to the world of medicine determine whether or not the study in question is a "good" or a "bad" study?

Here are some basic questions to get you started:

1) How was the study conducted? Was it a clinical study with the steps laid out prior? Was it done on humans or animals, and were the results tracked prior to the release of the drug?

2) How large was the study? This can be one of the most critical questions to research prior to accepting any study as "law." By rule of thumb, the larger the patient group in the study, the better. The larger the study, the more subjects, genetic types, and overall people have been tested. If no one in a study of ten people has any side effects, can you really generalize those results over a population of 5,000? The odds of a side effect being low or a drug being well tested are much better with a large cohort of people to spread the drug reaction across.

3) What was the length of time for the clinical trial? Was the study for seven weeks or seven years? Many of the new drugs on the market can be questioned because we have not seen the effects of their reactions over time. Think about tracking people over the course of five years of their life verses tracking them over a ten- or twenty-year time period. If you were only to get the snapshot of the average person during their college experience, could you really get a true idea of who that person was? The analogy is, the longer the timeframe, the more history and information you have as to what to expect.

4) Control, Double-blind, Placebo controlled, Randomized; know what these words actually mean…

There are certain terms which you should be familiar with if you are going to read through a clinical trial document. I have outlined the most common below in order to give you a leg-up in the terminology area.

Adverse reaction: an unwanted side effect which has been caused by the administration of drugs.

ARM: A treatment group in a randomized trial; also known usually as a clinical site.

Baseline: Information gathered at the beginning of a study which acts as a starting point to be compared to over time. This is the initial reference point that the changes in the patients or drugs are compared.

Blind study or trial: The patient or participant is not aware of whether or not he or she is getting the placebo, the drug, or something else.

Clinical investigator: The medical researcher who is responsible for carrying out the clinical trial's protocol.

Cohort: A group of individuals with some common characteristics.

Contraindication: A specific circumstance or patient type where the use of a certain drug or ingredient make-up can be harmful.

Control: The group or the medicine which is considered the comparison.

Diagnostic trials: Trials which are conducted in order to find better tests or procedures for diagnosis of a particular disease or condition.

Double-Blind study: A clinical trial where neither the participants nor the study staff knows which medicine or protocol the participants are receiving. This tends to be the well-respected method of drug studies; the thought process is that it gives the best objective results.

Drug-to-drug interaction: A change in the effect of a drug when administered with another drug. The effect may be an increase or a decrease in the action of either substance, or it may cause an adverse effect which is not normally associated with either drug.

Efficacy (of a drug or treatment): The ability of a drug or treatment to produce a result regardless of dosage. A drug passes efficacy trials if it is effective at the tested dose and is effective against the illness for which it is prescribed. Under FDA guidelines, Phase II clinical trials gauge a drug's efficacy, and Phase III trials confirm it.

Eligibility criteria: The criteria required for participant selection; this includes inclusion and exclusion criteria.

Empirical: A decision based on experimental data, not just on a theory.

Endpoint: The overall outcome which the drug or medical protocol is designed to evaluate. Common endpoints include severe toxicity, disease progression, or death.

Epidemiology: The branch of medical science that studies the incidence, distribution, and control of a disease in a population.

Experimental drug: A drug which is not approved by the FDA for use in humans, or as a treatment for a particular condition for which it is being tested.

Food and Drug Administration (FDA): The U.S. Department of Health and Human Services agency which is responsible for ensuring the safety and effectiveness of all drugs, biologics, vaccines, and medical devices, including those used in the diagnosis, treatment, and prevention of diseases.

Hypothesis: An assumption or potential theory used as a basis for reasoning or argument, or as a guide to experimental investigation.

Inclusion/exclusion criteria: The medical or social standards which will determine whether a person is allowed to enter a clinical trial. These criteria can include factors such as age, gender, the type and stage of a disease, previous treatment history, and other medical conditions

Informed consent: The process of learning the key facts about a clinical trial before deciding whether or not to participate in the trial. This is a continuing process to help someone decide whether or not to participate; the doctors and nurses involved in the trial explain the details of the study.

Informed consent document: This document describes the rights of the study participants, and includes details about the study, such as its purpose, duration, required procedures, and key contacts. The participant then decides whether or not to sign the document based on risks and benefits. Informed consent is not a contract, and the participant may withdraw from the trial at any time.

Institutional Review Board (IRB): 1. A committee of physicians, statisticians, researchers, community advocates, and others which ensure that a clinical trial is ethical, and that the rights of study participants remain protected. All clinical trials in the U.S. must be approved by an IRB before they are allowed to begin. 2. Every institution which conducts or supports biomedical or behavioral research involving human participants must, by federal regulation, have an IRB which initially approves and periodically reviews the research in order to protect the rights of human participants.

Intent to treat: The analysis of clinical trial results which usually includes the data from participants in the groups to which they were randomized, even if they never actually received the treatment.

Intervention name: The generic name of the intervention being studied.

Interventions: Primary interventions being studied: these types of interventions can include Drug, Gene Transfer, Vaccine, Behavior, Device, or Procedure.

Investigational new drug: An antibiotic drug, biological drug, or another new drug which is used in a clinical investigation. It also includes a biological product used in vitro for diagnostic purposes.

New drug application (NDA): An application submitted by the manufacturer of a drug to the FDA for a license to market the drug for a specified indication. The document gets submitted after the clinical trials have been completed.

Off-label use: A drug prescribed for conditions other than those approved by the FDA

Open-label trial: A clinical trial in which doctors and participants know which drug or vaccine is being administered. In other words, all parties involved are aware of which drug is being taken.

Orphan drugs: An FDA category which refers to medications used to treat diseases and conditions which rarely occur.

Peer review: The review of a clinical trial by experts chosen by the study sponsor. These experts review the trials for scientific merit, participant safety, and ethical considerations, and then submit the results back to the study sponsor.

Pharmacokinetics: The term used to describe the processes (in a living organism) of absorption, distribution, metabolism, and excretion of a drug or vaccine.

Phase I trial: The initial studies to determine the metabolism and pharmacologic actions of drugs in humans, the side effects associated with increasing doses, and to gain early evidence of effectiveness; may include healthy participants and/or patients.

Phase II trials: Controlled clinical studies conducted to evaluate the effectiveness of the drug for a particular indication or indications in patients with the disease or condition under study, and to determine the common short-term side effects and risks.

Phase III trials: Expanded controlled and uncontrolled trials after preliminary evidence suggesting effectiveness of the drug has been obtained, and are intended to gather additional information to evaluate the overall benefit-risk relationship of the drug and provide an adequate basis for physician labeling.

Phase IV trials: Post-marketing studies to delineate additional information including the drug's risks, benefits, and optimal use.

Placebo: An inactive pill, liquid, or powder which has no treatment value, in other words has no ability to treat the condition. This is usually a sugar pill or a derivative of something without any medication. In clinical trials, experimental treatments are often compared with placebos to assess the treatment's effectiveness.

Placebo controlled study: A method of investigation of drugs in which a substance with no treatment value (the placebo) is given to one group of participants, while the drug being tested is given to another group. The results obtained in the two groups are then compared to see if the investigational treatment is more effective in treating the condition.

Placebo effect: A potential physical or emotional change, occurring after a substance is taken or administered, which is not the result of any special property of the substance. The change may be beneficial, reflecting the expectations of the participant and often the expectations of the person giving the substance. This can be seen in participants who experience an effect which is not due to the drug's effect.

Preclinical: Refers to the testing of experimental drugs in the test tube or in animals prior to the testing of the drug on humans.

Prevention trials: Refers to trials to find better ways to prevent disease in people who have never had the disease, or to prevent a disease from returning. This can include medicines, vitamins, vaccines, minerals, or lifestyle changes.

Protocol: A plan on which all clinical trials are based. A protocol describes what types of people may participate in the trial; the schedule of tests, procedures, medications, and dosages; and the length of the study.

Randomization: A method based on chance by which study participants are assigned to a treatment group. Randomization minimizes the differences among groups by equally placing people with particular characteristics among all the trial arms. Therefore the researchers do not know which treatment is better.

Randomized trial: A study in which participants are randomly (i.e., by chance) assigned to one of two or more treatment arms of a clinical trial.

Risk-benefit ratio: The risk placed on individual participants versus the potential benefits. The risk/benefit ratio can differ depending on the condition being treated.

Screening trials: Refers to trials which test the best way to detect certain diseases or health conditions.

Side effects: Any undesired actions or effects of a drug or treatment. This description may include headache, nausea, hair loss, skin irritation, or other physical problems.

Single-blind study: A study in which one party, either the investigator or participant, is unaware of what medication the participant is taking.

Statistical significance: The probability that an event or difference occurred just by chance alone. In clinical trials, the level of statistical significance depends on the number of participants studied and the observations made, as well as the magnitude of differences observed.

Study endpoint: A primary or secondary outcome used to judge the effectiveness of a treatment.

Study type: The primary investigative techniques used in a study; the types can include Purpose, Duration, Selection, and Timing.

Toxicity: An adverse effect produced by a drug which is detrimental to the participant's health. The level of toxicity associated with a drug will vary depending on the condition which the drug is used to treat.

Treatment trials: Refers to trials which test new treatments, new combinations of drugs, or new approaches to surgery or radiation therapy.

5) Where was the research conducted? Was the research done in one site? Many sites? A hospital? A lab? Was it a large clinic, a university hospital, or some off-site facility on the outskirts of Guam? It is important to know who signed off on the trial, and where the research was actually done and completed.

6) Who paid for the study? As a representative selling the products in these trials, we were always taught to look at the back page of the study in question. There, at the end, all of the financiers are listed, and any pharmaceutical or medical company who funded or participated is listed. Here too is where any disclaimers will be stated if no funds exchanged hands during the trial. As a caution, this is an important place to look, but most clinical trials (due to their expense) have been paid for by medical or pharmaceutical companies. It is more important to see what the company influence was monetarily, and to see if they were the only financier or if there were others involved in the clinical trial funding.

7) Where was the trial published? This can be critical in determining the "market worthiness" of a study. If the study was published in JAMA or the New England Journal of Medicine, or any other major medical periodical, you can usually be assured that it was a worthy piece which the medical community took note of. There are other periodicals nationally and internationally which are

held with high acclaim in certain specialties, and any brief question to your physician or a Google search should give you the good vs. bad journal information. Pay attention to what journal the study was published in. If the journal was some off-shoot magazine that most physicians do not even read or consider serious, the study might be worth taking a more stringent look at. Sometimes clinical trials are forced to publication in order to "get them to market quicker." This can be a red flag if the journal is not well respected.

Reading clinical trials is at times considered to be boring, and takes a detailed eye to know what you are looking for. Just like learning a new vernacular in any native tongue, once you get the hang of it you will know where to go in the study to get the information which you need.

For the less faint of heart, I have detailed a basic clinical trial, what you can look for, and how to "cheat" without having to read an entire scientific study in order to get to the bottom line information. For those of you who were in remedial chemistry class, hopefully this will not be too daunting a task to delve further into the education of this subject.

There really is no "average" clinical trial paper to pull information off of, but most trials which are published follow a similar format.

Clinical trial 101: The first thing you will see on any trial paper is the journal name in which the study was published, the title of the paper, and the authors on the subject. If you really want in-depth information, look into the first or second person; find out their background and what kind of authority they have been on that topic before.

The next part of the study is my favorite. It is basically the abstract of the entire paper. This is, in essence, your cliffs notes version of what the study was about, its outcomes, population, etc. If you do nothing else, read the abstract in order to get the information about the trial, the side effects, and patients. Most of the time the abstract will follow this format:

Backgound: What drug was studied, why the trial was performed, and what the researchers were looking for.

Purpose: Why the study was done and what the hoped-for outcome was (if there was one).

Methods: How the researchers conducted their research. This will be an overview of some of the terms which I mentioned earlier: double-blind, placebo-controlled, and what sites were doing the research. Also, the methods category will give you the timeframe of the study, post-market information, and what the population was which was studied. This paragraph will also give you information about how many patients participated, what the population looked like, and the different study sites that were developed.

Results: What was the ultimate outcome? Did the researchers meet their initial goals? What was the side effect result, cure rate, retention rate, and follow up? This paragraph is great overall on what was actually found at the different time points on the study.

Conclusion: This is basically the summary paragraph which details the entire study reason, result, and outcome. A good question to ask is whether the researchers met their goals, outcomes, and results, from the purpose topic line.

If you read nothing else during the course of your time with the study, the abstract is a must. From this point, the study basically "fleshes" out the information further as to what the abstract stated. As I mentioned earlier it is like reading the cliffs notes version to Homer's *Iliad* and then taking the time to actually read the epic novel. The abstract will give you the basic overview. If you want to dig deeper, the novel is waiting.

Other high points to look at include the way that you look at the charts which are in the paper. Many times the authors will put charts and graphs of the important information, and one does not have to be a science student to see the value in the results.

Lastly, the side effect chart is usually on one of the back pages toward the end of the study. This is an important page to take a look at to see how the drug performed overall. As a caution, everything which occurs to any of the patients during the trial has to be listed on the side effect chart, so if you see something that one patient experienced which does not make sense for the drug's indication, it probably is an outlier from the overall drug effect. This is an important chart, though and can be critical to whether or not your comfort level is there short and long term.

As a sales representative I was always taught to go to the end of the study and look at the results and conclusions paragraphs. Even though most of this information was in the abstract, the results and conclusions paragraphs may have some additional information which was not listed. If you missed the entire *Iliad*, you at least would want to know how the story ended…right????

CHAPTER NINTEEN:

David Searls
Seeing death as the end of life is like seeing the horizon as the end of the ocean.

Saving Money on your Souvenirs and Travels:

So you take your time, do your educational study, and finally have ventured into the world of medicine. The shock factor of your trip comes when you are charged a LOT of money for the souvenirs or medication which is prescribed. Medicines can give the average patient quite a large setback financially when they are down with their visit to the physician's office or hospital.

As a representative, I learned the differences between some of the drug plans on the market and a couple of easy tricks of the trade to save a few dollars here and there.

According to many statistics, there are over 3 billion prescriptions per year. That is a lot of souvenir money on the table!!

43-45 percent of Americans receive at least one prescription per month; many patients are receiving 2-3 prescriptions, and many times they are coming from different doctors.

Here are some of the things which might help as you navigate through the souvenir shops on your travels:

Whenever possible, try to use just one pharmacist or pharmacy. Not only can this person look at the different medications which you are taking, but he or she can also educate you about the drugs which are being prescribed. A good pharmacist is a true friend to have. He can advise you about possible risks, look at the potential for drug-to-drug interactions with the medications that you are taking, and help you to understand your insurance benefits. Pharmacists are usually able to track your prescriptions through their computer system, and a "red flag" will pop up if something is prescribed which might interact with your other medications.

If you are seeing more than one physician for your care, your pharmacist may just be the intermediary that you need to determine and interpret the differences between the souvenirs which you are being prescribed.

There are many, many options in the world of souvenirs as to how to pay for and potentially save money on your pharmaceutical prescriptions. The problem is that there are so many options out there that very few people have the time to read and understand all of the information in order to make a truly educated decision.

I am briefly going to describe the different options available to the average consumer. Additionally, www.aarp.org, www.medi care.gov, and most of the large pharmaceutical/medical companies will highlight any additional details that you might need.

People who are a part of a private medical insurance plan usually have a drug benefit as a part of their insurance policy. This plan usually covers drug prescriptions on a tiered basis. The formulary is the term for a short list of drugs that is voted onto a hospital or pharmacy's list which becomes the "preferred list" for doctors and pharmacists to prescribe and dispense. Usually the formulary is chosen by the Pharmacy and Therapeutics Committee (P&T Committee) of an institution or hospital, and drugs are placed there based on cost incentives, efficacy parameters, and honestly, likes and dislikes of the consensus of the committee members.

Drugs can be on a formulary on a tiered basis, where the first choice product will cost you a smaller amount of money for a copay than the third choice might.

Beware that some products may not be covered by the formulary, such as lifestyle drugs (diet pills, anti-depressants, sexual enhancement products). It honestly depends on the committee and their decision-making processes.

Many times the tiered level can be based on generics vs. branded formulations of the drug. A generic drug is usually a similar make-up of the original branded formulation, but the recipe of the drug has been released and re-branded due to time on the market. As a result, the generic drug cost is usually less.

Doctors and pharmacists are sometimes graded by what products they are prescribing. They can be held to the formulary by fines or graded evaluations if they move off of the formulary drugs on a regular basis. It is important that you know where your drug of choice stands so you can request the most economic but effective product for your needs.

Generics vs. branded drugs:

The definition of generics vs. branded products is best explained as souvenirs at the mini-mart vs. the high-end retail shops that you might see while on vacation. When a drug initially is brought out to market it usually has a branded name, a premium price, and a certain patent timeframe which has been given to the drug company to hold on to their recipe. Once that timeframe has passed, any company can buy the rights to the recipe and create a generic version of the product. Many times the original drug company will even drop its own product to a generic price in order to earn the most they can once a drug's ingredient make-up has been released.

The rules for a generic vs. a branded drug are that the active ingredients of the generic product need to be the same as the branded drug. The inactive ingredients can vary, but usually a certain percentage of the product's formulation has to remain the same. You may want to clarify with your pharmacist about the generic medication which you are choosing, as to what exactly the difference is of that specific medication. Although usually it is no big difference with regard to effectiveness, it never hurts to understand exactly what the trade-off is for your saving some money.

Therapeutic substitutions also are sometimes recommended at the pharmacy counter. These are drugs which are in a class similar to the branded medication, but are not required to have the same active or inactive ingredients as the original product. This is important to clarify the exact differences, especially if the therapeutic or generic is being suggested for a serious medical condition. You always have the right to have your physician write an appeal to your insurance company or pharmacy in order to insure that you are getting the drug you want.

There are some additional ways to save money on your souvenirs in the world of medicine. If your physician is instrumental in the formulary process, you may want to see about their influence within the next vote of the committee to get your drug of choice either added or moved up on the list of preferred products.

Physicians are showered with samples from pharmaceutical representatives. Explain your plight to your physician and pharmacist; many times they have plenty of samples to spare which have not cost them a dime, but can save you a few dimes.

Be sure to research the product you are taking on the medical or pharmaceutical company's web site. Many companies are now offering drug assistance plans and policies, joint drug benefit plans, and rebate coupons based on the product you choose.

Canada and online drug ordering has become an option for many due to high drug costs, and without delving too deeply into the legality and safety of this option, I just urge caution if you choose to go the online route. Using our travel analogy, if you wanted to save money on an authentic souvenir, you could try to find it online at a cheaper web site, but without actually seeing the natives who made it, or buying it in person, you do run the risk of ending up with an imposter product rather than the real thing.

Just like anything else, it pays to shop around when looking at your souvenir options in the world of medicine. If you buy the souvenir at the first store you come to in a foreign land, the natives might be taking advantage of a prime tourist spot. Around the corner you may find the same product for a bargain price. Use the experts (doctors, pharmacists, and pharmaceutical reps) to your advantage to ensure that you are getting the best authentic product for the best price possible.

Avoiding outrageous hospital overcharges is another educational lesson that I learned about as a medical representative.

As an uneducated traveler though, how do you protect yourself when traveling into the world of medicine?

How can you avoid getting stuck with unnecessary charges, or being overcharged on your bill?

There are a few critical steps which you can take to ensure your stay is covered and that you are protected from the issues of being overcharged:

Try to find out in advance what your medical plan covers and what it will pay. Attempt to be as educated as possible about the details of how your coverage works and what you are actually covered for. If you end up in an emergency situation you may have to trust some of these details to the institution, but if you have time to plan, you can take the time to ensure that you are covered.

Get to know the hospital billing department if you can. A simple phone call to ask what your estimated charges will be is a good idea. Just because it is a hospital does not mean that you should not treat this as if you were staying in a hotel. If there are "extras" (i.e. tissues, toilet paper, food items) on the bill, see if you can bring your own items in advance in order to save on your overall visit.

Make sure that all of the physicians and specialists who will be in charge of your care are covered by your medical plan. Be sure to double check with the hospital and staff about any special charges which might not be covered by your medical plan or insurance.

Keep detailed notes of tests, medications, and treatments. You should be able to compare your treatments, medications, etc., to the itemized bill which you will ask for from the billing department. Get as much additional information from your doctor about your treatment, average costs, etc. He should be able to tell you about most of the tests, operational procedures, etc., and ballpark costs.

Read and re-read all documentation, bills, notes, and any other written material which you receive from your medical plan or insurance company regarding your hospital stay.

Never pay your bill before leaving the hospital. If you need to consult with an attorney or patient advocate, they will tell you that paying prior to leaving is not a requirement. Even if you are told you need to pay prior to departure, you can refuse based on your patient rights.

After you ask for an itemized bill from the hospital, compare the details to those in the notes which you made during your initial visits and see if any differences have occurred. Do not accept "miscellaneous" as a category on any bill. Also, you should not accept anything that you do not understand. Mistakes do get made periodically on these bills, and each category should be well-explained prior to you paying for it.

If you come across something which is incorrect, or you are unsure of its meaning, bring it to the attention of a hospital administrator, a patient advocate, or a patient ombudsman. These groups are there for your protection. You should be able to obtain a complete explanation of your bill and medical information prior to paying for any of the additional details.

In medical sales I was always amazed that patients were willing to take the word of a hospital or physician and pay for a bill or for medication without looking at the fine print or for a cheaper price. If you were taking a trip to a foreign land, you would be very wary of hotel pricing, rates, food costs, etc. A trip into medical world should be no different. You should have the opportunity to scrutinize each and every line item to insure that you are getting the best overall pricing and care which you deserve.

CHAPTER TWENTY:

Elizabeth Kubler-Ross
Learn to get in touch with silence within yourself, and know that everything in this life has purpose. There are no mistakes, no co-incidences, all events are blessings given to us to learn from.

Your Rights in and Beyond Medical World: Patients' Rights and Responsibilities/HIPAA:

If you were going off to a foreign country, you would probably want to know your rights as an American, or citizen of another country, prior to travel, right? The world of medicine is no different.

You have been given certain rights and responsibilities in your visit into the medical world. The following is a shortened version of the actual bill of rights which was created by the American Hospital Association in 1973.

1) You have the right to informed consent in treatment decisions, access to specialty care, and confidentiality of your medical records and communication with providers.
2) You have the right to receive concise and easy to understand information about your medical coverage.
3) You have the right to know how your coverage decisions are made and how they can be appealed.
4) You have the right to receive information about the cost of your coverage and care.

5) You have the right to a reasonable choice of providers and provider options when seeking medical care.

6) You have the right to know about provider incentives and restrictions in the influence of patient care or practice patterns.

7) You have the right to be responsible about the way in which you take care of your body, which includes eating properly, exercising, sleeping well, and practice of stress management.

8) You have the right to education and knowledge about your medical care, health plans, and physician facilities.

9) You have the right to be involved in your own treatment and care.

10) You have the right to ask for a second opinion or opinions if something does not sound correct.

If you do your research, you will realize that there are many other lists of "rights" out there which offer similar information for your review.

Additional rights were outlined by the Advisory Commission on Consumer Protection and Quality in the Health Care Industry under the Clinton Administration.

Those, too, are outlined below for your information:

Patients' Bill of Rights

I. Information Disclosure

You have the right to receive accurate and easily understood information about your health plan, health care professionals, and health care facilities. If you speak another language, have a physical or mental disability, or just don't understand something, assistance will be provided so you can make informed health care decisions.

II. Choice of Providers and Plans

You have the right to a choice of health care providers that is sufficient to provide you with access to appropriate high-quality health care.

III. Access to Emergency Services

If you have severe pain, an injury, or sudden illness which convinces you that your health is in serious jeopardy, you have the right to receive screening and stabilization emergency services whenever and wherever needed without prior authorization or financial penalty.

IV. Participation in Treatment Decisions

You have the right to know all your treatment options and to participate in decisions about your care. Parents, guardians, family members, or other individuals who you designate can represent you if you cannot make your own decisions.

V. Respect and Nondiscrimination

You have a right to considerate, respectful, and nondiscriminatory care from your doctors, health plan representatives, and other health care providers.

VI. Confidentiality of Health Information

You have the right to talk in confidence with health care providers and to have your health care information protected. You also have the right to review and copy your own medical record, and request that your physician amend your record if it is not accurate, relevant, or complete.

VII. Complaints and Appeals

You have the right to a fair, fast, and objective review of any compliant you have against your health plan, doctors, hospitals or other health care personnel. This includes complaints about waiting times, operating hours, the conduct of health care personnel, and the adequacy of health care facilities.

Now that I have given you your rights, what exactly does this mean and how do you exercise these when visiting the world of medicine? This can be a bit confusing to read about all of your rights prior to arrival, but how do you protect yourself in the moment?

Most of us know the rules and laws of driving, right? Many of us break them unwillingly when suddenly we are forced into a tough situation on the road which requires emergency action.

The world of medicine is no different. You might be in a difficult state of mind upon arrival, and will not have the opportunity to remember all of your rights in the midst of medical confusion.

Here are some of the things that you can do pre- and post-visit to protect yourself at all times.

You can request a release of your medical records and can keep them for your own information at home or the office. These records can be amended if they have any incorrect information, and you should review them periodically much like you would your credit report or driving record.

Additionally, you will be privy to the hospital or medical facilities' privacy practices. If in any way your privacy is breached, you can file a complaint with the Department of Health and Human Services by going to www.hhs.gov/ocr/hipaa or calling 800-368-1019. HIPAA is the law which was passed in April of 2003. HIPAA stands for the Health Insurance Portability and Accountability Act, which basically protects your rights when it comes to your medical records.

You are guaranteed by HIPAA to look at and request changes to your health and medical information, and this basically gives you, the patient, more control over who gets access to your records from medical care.

In addition to HIPAA, each state has its own rules as to what a patient's rights should be. It is worth doing a bit of homework into your states' rules and regulations prior to your visit to the world of medicine.

So, you have all of your rights, what else do you need?

You need to be aware of your records, your medical history, and what your wants are in the event that you do not return in the same condition from the world of medicine. Although no one likes to think about the unfortunate things which can occur at times in life, in the medical world the best protection that you have is to not only understand your rights and document everything, but also have the best documentation possible in case of emergency.

Whenever I travel on vacation or business, I call my relatives and tell them briefly where all of my financial assets are, "just in case". This should be the same thought process that one should take when visiting the world of medicine.

Some of the documents which you might want to have at your disposal prior to your trip are:

Your medical records

A Living Will

Power of Attorney (Healthcare or Medical)

Do Not Resuscitate Orders

These can be prepared fairly easily by an attorney, if necessary, and there are many on-line programs and books which describe each and give a template or background about what they need to be effective.

Every time I take a trip, I make sure that everything I need (passport, guidebook information, plane tickets, insurance, etc.) is located where I can find it and is easily accessible. The world of medicine should be no different.

Get your documentation together. Make sure you are protected, before you get set to board the vessel into your new world.

As I mentioned earlier, a medical or healthcare advocate can also be helpful in getting this information together. As an aside, the trend in both medical practices and hospitals today is to move toward electronic medical records. This has created a bit of a mess for all facilities as they try to get outdated software and medical records and convert them into present-day electronics and beyond.

You may want to ask the facility which you are about to visit what their preference is for record information. If they are electronic, you can bring a small "thumb" drive storage device (any office supply store should have these), and carry all of your information with you on the small driver device for ease of access for the clinic or hospital.

Online records are becoming a trendy thing to discuss as hospitals and clinics are moving into the information age more readily. Suddenly there has been discussion by the U.S. Department of Health and Human Services to create a National Health Information Network in order to completely digitized medical information of all Americans.

This sounds like a great plan, and many public and private investors are interested in creating a database like this sooner rather than later. Unfortunately, though, we are a long way off from any test site for this kind of an overhaul of medicine. So while working within the confines of the current system, it is up to each individual to protect his or herself by keeping, and updating, the information which we discussed earlier.

CHAPTER TWENTY-ONE:

Unknown
Life may not be the party we hoped for, but while we're here we should dance.

Traveler's Insurance: Insurance 101 (Medicare, Medicaid, PPOS, HMO's, oh my!):

When traveling to a foreign country, many guidebooks suggest that you take out some medical insurance "just in case." This should be mandatory in any visit into the world of medicine.

There are many different options out there, and although they may sound expensive or look confusing at the beginning, once you take a closer look it can be fairly easy to choose a protective option with confidence.

Again, as mentioned earlier, entering into the world of medicine takes some planning and preparation. You want to be ready for the "just in case" scenarios.

Here is some background on the different options available:

HMO: An HMO (Health Maintenance Organization) refers to an organization which provides for coverage of certain health care services required by members of that group. Typical HMO coverage can include access to a primary care physician, emergency care, and specialists/hospitalization as necessary; it is important to read the fine print of your HMO plan to ensure that you are aware of what is covered and any additional items which are not.

Some of the critical opposition to HMO's addresses concerns as to a lack of selection of primary care physicians. There have been stories of denials of adequate referrals in the event of disease or illness. Critics have claimed that HMO's may deny certain claims and may make health care decisions based upon a pure profitability standpoint, as opposed to decisions driven by providing the best level of care for its patients. HMO's generally do a good job of tracking patient outcomes, overall care rates, and managing patient economics. Although sometimes they are accused of choosing for the economic care of the HMO rather than the overall concern of the patient.

IPO: An IPO, which stands for Independent Provider Organization, operates by linking an HMO contract with independent physicians to provide services to HMO members.

PPO: PPO, or a Preferred Provider Organization, is a form of managed care under which health care providers contract to provide medical services at pre-negotiated rates. Members who subscribe to a PPO are required to use the health care providers who participate in the PPO network or are usually charged an out-of-pocket fee or penalty if they go outside of the network.

POS: POS stands for Point of Service plans. These plans allow the individual policy holder or certificate holder to visit out-of-network, non-participating doctors for an additional fee. If the services of a non-participating health care provider are utilized, the individual often obtains restrictions of benefits or incurs more out-of-pocket costs than they would have if they had stayed within the network.

Medicare and Medicaid also are important for the average consumer to understand. There has been a lot of news in recent years regarding Medicare. The new Medicare coverage plan was introduced in January 2006 by President George W. Bush. Here are some of the basics in defining the who, what, why, and how of the Medicare process:

Medicare is a health insurance program for:

- people age 65 or older;
- people under age 65 with certain disabilities; and

- people of all ages with End-Stage Renal Disease (a.k.a. permanent kidney failure requiring dialysis or a kidney transplant).

Medicare has a few different options:

Part A Hospital Insurance - Most people don't pay a premium for Part A because they or a spouse already paid for it through their payroll taxes while working. Medicare Part A, which falls under Hospital Insurance, helps cover inpatient care in hospitals, including critical access hospitals, and skilled nursing facilities. It also will help to cover hospice care and some home health care.

Part B Medical Insurance - Most people pay a monthly premium for Part B. Medicare Part B, which is medical insurance, helps cover doctors' services and outpatient care. It also covers some other medical services which Part A doesn't cover, such as some of the services of physical and occupational therapists, and some home health care. Part B helps pay for these covered services and supplies when they are medically necessary. Again, beneficiaries must qualify in order to be covered.

Prescription Drug Coverage - Most people will pay a monthly premium for this coverage. Starting January 1, 2006, new Medicare prescription drug coverage was made available to everyone with Medicare. This means that everyone with Medicare can get this coverage, which may help lower prescription drug costs and help protect against higher costs in the future. Medicare Prescription Drug Coverage is basically just insurance. Private companies are responsible for providing the coverage. Beneficiaries choose their drug plan and then pay a monthly premium. Like other insurance benefits, if a beneficiary decides not to enroll in a drug plan when they are first eligible, they may pay a penalty if they choose to join later.

As you can see, there is a lot to know about Medicare and the differences between the types available. It can be really confusing to read through the governmental words and try to understand the different coverage and how they protect the average consumer. Hopefully this guide is a useful differentiator between the different options out there.

Medicaid, on the other hand, is available only to certain low-income individuals and families who fit into an eligibility group which is recognized by Federal and State law. Medicaid, in comparison, does not pay money to you; instead it sends payments directly to your health care providers. Depending on your state's rules, you may also be asked to pay a small part of the cost for some medical services.

Medicaid is a state-administered program, so each state is responsible for setting its own guidelines regarding eligibility and services of Medicaid. If you are unhappy with the Medicaid in your state, it may be time to move…just kidding.

Certain requirements or criteria usually must be met under Medicaid guidelines. These requirements can include age, pregnancy, whether you are disabled, blind, or aged; your income level and/or additional resources (like bank accounts, real property, or other items which can be sold for cash). Other stipulations can include whether you are a U.S. citizen or a lawfully admitted immigrant. The rules for counting your income and resources vary from state to state and from group to group. It is important that you learn the rules of your specific state prior to trying to claim anything through the Medicaid system. It will save you a lot of time and energy.

Your child may be eligible for coverage if he or she is a U.S. citizen or a lawfully admitted immigrant. Child laws can vary depending on their status as a citizen, their guardian status, and their age. If your income is low and you match one of the descriptions of the eligibility groups, you probably can apply to receive Medicaid.

Whew! What an exhausting journey through the defining factors behind the Medicare and Medicaid systems! Onward, weary traveler, we are almost done.

CHAPTER TWENTY-TWO
(the additional chapter)

Mickey Mantle
If I knew I was going to live this long, I'd have taken better care of myself.

It is only fair to mention that this book was originally only going to be twenty-one chapters long. When I wrote the Travelers Guidebook to the World of Medicine, I ended it the first time right at chapter twenty-one.

As I completed the initial rough draft, I called up my grandfather, who (for those who remember the introduction) was the true reason behind this book.

My Papa asked me politely for a copy of the manuscript, and being a good granddaughter, I sent it over via e-mail a bit sheepishly.

Knowing the scientific mind and background that my grandfather has, I knew that when I did not hear back immediately on my draft, it only meant he was reading my writing line-by-line with a careful eye and a discerning mind.

My pulse quickened a bit and my blood pressure rose, but I had given the document over willingly and was ready to hear how good or bad it was from start to finish.

As expected, the feedback which I received from the returned e-mail was not necessarily good or bad, but sounded rather perplexed in nature. My grandfather liked the ideas in the book, thought the information was good, but wanted even more ideas for the readers regarding how to make the world of medicine a "more efficient" experience. He felt that a lot of his time and other patients' energy had been wasted in the doctors' offices and hospitals. My Papa thought time and energy could be used more effectively with better technology and a more educated patient base.

So, after much thought and discussion, this final chapter was written to describe how you, the traveling patient, can make your visit an even more efficient one in the world of medicine.

After all the education of the previous chapters, the only critical piece that seems to be missing is a tie-in of technology or a how-to guide on more efficient processes in the world of medicine.

I do agree with my Papa that we have a long way to go. The medical industry is changing, but changing too slowly in my opinion, to offer a true technology solution in the short-term.

After my experience as a representative, I feel the best way for a patient to truly improve the efficiency of a medical practice is to use the knowledge about his or herself to dictate better patient care and outcomes.

There are many technology companies today who are creating electronic tools for medical records and patient histories. If each person would just be savvy enough to keep a transcript of their information, insurance policies, pharmaceuticals, and overall health history in a diary form, on a memory stick, or on an electronic record, millions of dollars in medical mistakes could be saved immediately.

My comments back to my grandfather and my position with you, the traveling patient, is that the industry itself is changing, but the real change has to start first with us. We cannot expect each hospital, medical office, family practice doctor, and/or the medical system to ramp up their technological knowledge and computer expertise overnight. We cannot expect the federal government to dig further into our own pockets to analyze and overhaul the entire medical industry before the baby boomers all hit retirement. We cannot expect that everyone will be treated without medical mistakes in every circumstance today as the system in place evolves.

What we can expect is that with patient knowledge comes power, with patient understanding comes compassion, and with common communication between practitioner and patient comes better care. Thus far I have not seen enough change in the short period of time to warrant a dramatic difference in patient care in the timeframe in which we need it to adjust.

Many patients feel as if they are still traveling to a foreign world, and the system is not catching up fast enough to protect the average Joe or Jane.

My final comment to my grandfather was that I had spent a decade trying to educate and change the medical world from within as a medical representative. After many attempts, my final resolution became to educate and change the medical world from outside, by appealing to the you, "traveling patient".

Just as if you traveled to a foreign world as a missionary, you would quickly become aware that you alone will not change the customs, laws, and systems of that world overnight. It is only by living among the natives, communicating in their tongue, and learning their traditions that you will see changes over the course of time; Both in you, as well as in the locals.

With that conclusion, I officially hand you your medical passport to the world of hospitals and physicians, wish you safe travels, and ask that you do your best to make my grandfather proud of the next generation as we travel into the world of medicine with less effort and a more efficient purpose.

GLOSSARY:

Dictionary of Medical Terms and Phrases in the World of Medicine:

You can find many of the best terms for medical purposes at www.medterms.com, or www.webmd.com, or by catching the latest episode of ER. Just like any other guidebook, however, here are some glossary phrases to get you started in your translation with the natives in the world of medicine.

Abdomen - area between the chest and the hips that contains the stomach, small intestine, large intestine, liver, gallbladder, pancreas, and spleen.

Abdominal hysterectomy - the uterus is removed through the abdomen via a surgical incision.

Abdominoscopy - a type of surgery which uses a laparoscope, which is inserted into one or more small incisions, to examine the abdominal cavity.

Absorption - the way substances pass into tissue as nutrients from food move from the small intestine into the cells of the body.

Accessory movement - joint movements which cannot be performed voluntarily or in isolation by the patient.

Accommodation - the ability of the eye to focus.

Acetylcholine - a chemical in the brain which acts as a neurotransmitter by sending nerve signals.

Acquired deafness - loss of hearing which occurs or develops over the course of a lifetime; deafness not present at birth.

Activities of daily living (ADL's) - personal care activities necessary for everyday living, such as eating, bathing, grooming, dressing, and toileting; a term often used by healthcare professionals to assess the need and/or type of care a person may require.

Acute - refers to a medical problem which requires immediate attention. In other words, if a patient has an acute problem or a matter is acute, it is immediate and needs attention ASAP.

Ad Lib: Freedom (usually by the patient) to do what they want (i.e., the patient can now resume regular physical activity ad lib).

Adenocarcinoma - a cancer which develops in the lining or inner surface of an organ.

Adjuvant treatment - treatment which is added to other therapies to increase effectiveness, such as chemotherapy, radiation therapy, or hormone therapy.

Adrenal cortex - the outer portion of the adrenal gland that secretes hormones which are vital to the body.

Advanced cancer - stage of cancer in which the disease has spread from the primary site to other parts of the body; when the cancer has spread only to the surrounding areas, it is called locally advanced; when it has spread further by traveling through the bloodstream.

Aerophagia - condition which occurs when a person swallows too much air; causes gas and frequent belching.

Agonist - a drug capable of combining with receptors to initiate an action that can be known in advance.

-Algia: usually at the end of the word, it refers to pain or ache

Alimentary canal - digestive tube.

Allergy: the reaction of the patient's immune system to pollen, dust, animals, certain foods, or other substances.

Allogeneic bone marrow transplantation - a procedure in which a person receives stem cells from a compatible donor.

Alternative medicine - any form of therapy used alone, without recommended standard/conventional treatment.

Amenorrhea - absence or cessation of menstrual periods.

American Cancer Society - an organization which supports research, educational materials, and programs, and offers many other services to cancer patients and their families.

American Sign Language (ASL) - manual (hand) language with its own syntax and grammar, used primarily by persons who are deaf.

AMI: Acute Myocardial Infarction; a heart attack. If a patient is having an acute myocardial infarction, a cardiologist is on the way.

Amyloidosis - a rare disease which causes the build-up of amyloid, a protein and starch, in tissues and organs.

Anal fistula - channel which develops between the anus and the skin. Most fistulas are the result of an abscess (infection) which spreads to the skin.

Androgen - a hormone which promotes the development and maintenance of male sex characteristics.

Anesthesia - loss of feeling or sensation as a result of drugs or gases. General anesthesia causes loss of consciousness; local or regional anesthesia causes loss of feeling only to a specified area.

Anesthetics - drugs which cause loss of sensation to pain or awareness.

Angiotensin - converting enzyme (ACE) inhibitor - a medication which lowers blood pressure.

Ankle sprain - overstretched lateral (outside) ligament of the ankle joint.

Antacids - medicines which balance acids and gas in the stomach.

Anterior chamber of the eye - the front section of the eye's interior where aqueous humor flows in and out of, providing nourishment to the eye and surrounding tissues.

Antibiotic - chemical substances produced by living organisms or synthesized (created) in laboratories, for the purpose of killing other organisms that cause disease.

Anticholinergics - medicines which calm muscle spasms in the intestine.

Anticoagulant (also called blood thinner.) - a medication which keeps blood from clotting.

Antidiarrheals - medicines which help control diarrhea.

Antiemetic - drug which prevents or relieves nausea and vomiting (emesis).

Antihistamine drugs - a group of drugs which block the effects of histamine, a chemical released in body fluids during an allergic reaction.

Antihypertensive - a medication, or other therapy, which lowers blood pressure.

Anti-inflammatory drugs - inflammatory drugs which reduce the signs and symptoms of inflammation.

Antimetabolites - substances which interfere with the body's chemical processes, such as creating proteins, DNA, and other chemicals needed for cell growth and reproduction; in cancer treatment, antimetabolite drugs disrupt DNA production, which in turn prevents cell division.

Antioxidants - compounds which protect against cell damage inflicted by molecules called oxygen-free radicals, which are a major cause of disease and aging.

Antispasmodics - medicines which help reduce or stop muscle spasms.

Antrectomy - operation to remove the upper portion of the stomach, called the antrum, often to help reduce the amount of stomach acid.

Anus - opening at the end of the digestive tract where bowel contents leave the body.

Aorta - blood vessel which delivers oxygen-rich blood from the left ventricle to the body; it is the largest blood vessel in the body.

Aortic valve - the valve which regulates blood flow from the heart into the aorta.

Apex - top portion of the upper lobes of the lungs.

Aphasia - total or partial loss of ability to use or understand language; usually caused by stroke, brain disease, or injury.

Apraxia - inability to make a voluntary movement in spite of being able to demonstrate normal muscle function.

Areola - dark area of skin which surrounds the nipple of the breast.

Arterioles - small branches of arteries

Artery - a blood vessel that carries oxygenated blood away from the heart to the body.

Arthralgia - pain in a joint, usually due to arthritis or arthropathy.

Arthroplasty - total joint replacement.

Articular cartilage - covers the ends of bones and allows the distribution of compressive loads over the cross section of bones; provides frictionless and wear-resistant surface for joint movement.

Articulation disorder - inability to correctly produce speech sounds (phonemes) because of imprecise placement, timing, pressure, speed, or flow of movement of the lips, tongue, and/or throat.

Artificial ventilation - the process of supporting breathing by manual or mechanical means when normal breathing is inefficient or has stopped.

Assistive devices - technical tools and devices such as alphabet boards, text telephones, or text-to-speech conversion software used to assist people with physical or emotional disorders in performing certain actions, tasks, and activities.

Atherectomy - a non-surgical procedure which involves removing plaque from the walls of arteries with a rotating blade.

Atherosclerosis - a type of arteriosclerosis caused by a build-up of plaque in the inner lining of an artery.

Atrioventricular block - an interruption of the electrical signal between the atria and the ventricles.

Atrium (pl. atrial) - one of two upper chambers in the heart.

Atrophic gastritis - chronic inflammation of the stomach that causes the breakdown of the mucous membranes of the stomach.

Atrophy - a continuous decline of a body part or tissue, usually a muscle, following a period of disuse or immobility.

Atypical - not usual; abnormal; often refers to the appearance of pre-cancerous or cancerous cells.

Auditory brainstem response (ABR) test - test used to screen for hearing impairments in infants and young children.

Auditory nerve - eighth cranial nerve which connects the inner ear to the brainstem.

Auditory perception - ability to identify, interpret, and attach meaning to sound.

Auditory prosthesis - device which substitutes or enhances the ability to hear.

Augmentative devices - tools which help individuals with limited or absent speech to communicate.

Autoimmune process - a process where the body's immune system attacks and destroys body tissue that it mistakes for foreign matter.

Autopsy - examination of a body after death; performed to determine cause of death, or to verify a diagnosis.

Avascular necrosis - death of tissue due to a depletion of blood supply.

Axilla - armpit.

Axillary dissection - a surgical procedure in which the lymph nodes in the armpit (axillary nodes) are removed and examined; often used to determine if breast cancer has spread to the axillary nodes.

Axon - the long, hair-like extension of a nerve cell that carries a message to a nearby nerve cell.

Bacteria - the one-celled microorganisms found in air, water, and soil which can cause infectious diseases.

Balloon angioplasty (also called coronary artery balloon dilation.) - a procedure used to widen narrowed arteries. The procedure uses a catheter with a deflated balloon at the tip which is inserted into the narrowed part of the artery. The balloon is then inflated, causing the artery to dilate (widen).

Barium - a metallic, chemical, chalky, liquid used to coat the inside of organs so that they will show up on an x-ray.

Barium swallow - upper gastrointestinal (GI) series.

Barotrauma - injury to the middle ear caused by a reduction of air pressure.

Basal body temperature - temperature of a person's body taken first thing in the morning after several hours of sleep and before any activity, including getting out of bed or talking; often charted to determine the time of ovulation.

Basal ganglia - several large clusters of nerve cells, including the striatum and the substantia nigra, deep in the brain below the cerebral hemispheres.

Basal metabolic rate (BMR) - a measurement of energy required to keep the body functioning at rest. Measured in calories, metabolic rates increase with exertion, stress, fear, and/or illness.

Base of the lung - bottom portion of lower lobes of the lungs, located just above the diaphragm.

Benign - not harmful or cancerous, when referring to tumors.

Beta blocker - an antihypertensive medication which limits the activity of epinephrine (a hormone that increases blood pressure). A class of drugs used to treat high blood pressure by slowing the heart rate.

Beta cells - cells which make insulin, found in areas of the pancreas called the islets of Langerhans.

Biarthrodial muscles - muscles which span over two joints and have a function over those joints.

Bilateral - affecting both sides of the body; for example, bilateral breast cancer is cancer occurring in both breasts at the same time (synchronous) or at different times (metachronous).

Biliary tract (also called biliary system or biliary tree) - gallbladder and the bile ducts.

Binocular vision - the ability to use both eyes at once.

Biologic response modifiers - substances which boost the body's immune system to fight against cancer.

Biological therapy (also called immunotherapy, biotherapy, or biological response modifier therapy) - uses the body's immune system, either directly or indirectly, to fight cancer or to lessen side effects that may be caused by some cancer treatments.

Bladder - a hollow organ which urine drains into from the kidneys.

Blink rate - the number of times per minute that the eyelid automatically closes - normally 10 to 30 times per minute.

Blood plasma - the fluid part of blood which contains nutrients, glucose, proteins, minerals, enzymes, and other substances.

Blood pressure cuff - a device usually placed around the upper part of the arm to measure blood pressure.

Blood-brain barrier - the protective membrane which separates circulating blood from brain cells.

Body mass index (BMI) - a number, derived by using height and weight measurements, which gives a general indication of whether or not weight falls within a healthy range.

Bone - living tissue that makes up the body's skeleton.

Bone density test - a test which measures the strength and density of bones; often used to determine the risk of developing osteoporosis.

Borborygmi (also called stomach "growling") - rumbling sounds caused by gas moving through the intestines.

Botanical - of plants and plant life.

Bowel - another word for the small and large intestines.

Bowel movement - body wastes passed through the rectum and anus.

Bowel prep - process used to clean the colon with enemas and a special drink; used before surgery of the colon, colonoscopy, or barium x-ray and other imaging exams.

Brady... - suffix meaning slow.

Bradycardia - abnormally slow heartbeat.

Bradykinesia - slowness of movement.

Bradyphrenia - slowness of thought processes.

Brain attack - another name for stroke.

Brain scan - an imaging method used to find abnormalities in the brain, including brain cancer and cancer that has spread to the brain from other places in the body.

Brainstem implant - auditory prosthesis which bypasses the cochlea and auditory nerve to help individuals who cannot benefit from a cochlear implant because the auditory nerves are not working.

Braxton Hicks contractions - relatively brief, painless contractions of the uterus which may begin during the second half of pregnancy.

BRCA1 - a gene which, when damaged or mutated, places a woman at a much higher risk of developing breast and/or ovarian cancer than the general population.

BRCA2 - a gene which, when damaged or mutated, places a woman at a much higher risk for developing breast cancer and/or ovarian cancer than the general population.

Breast cancer - a cancerous tumor of the breast tissue.

Breast conservation therapy - surgery to remove a breast cancer and a small amount of benign tissue around the cancer without removing any other part of the breast. This procedure is also called lumpectomy, segmental excision, limited breast surgery, or tylectomy.

Breast implant - a manufactured sac that is filled with silicone gel (a synthetic material) or saline (sterile saltwater); the sac is surgically inserted to increase breast size or restore the contour of a breast after mastectomy.

Breast reconstruction - surgery to rebuild a breast mound after mastectomy.

Breast specialist - term describing health care professionals who have a dedicated interest in breast health. While they may acquire specialized knowledge in this area, medical licensing boards do not certify a specialty in breast care.

Bronchodilators - a group of drugs which widen the airways in the lungs.

Bronchus (pl. bronchi) - one of the two large airways of the lungs. The trachea divides into the right and left bronchi.

Bulking agents - laxatives which make bowel movements soft and easy to pass.

Bundle-branch block - a condition in which the heart's electrical system is unable to normally conduct the electrical signal.

CA-125 test - blood test to detect an elevated level of a protein antigen called CA-125, which may indicate ovarian cancer, among other disorders.

Calcium channel blocker (also called calcium blocker) - a medication used to treat heart disease.

Calculi - stones or solid lumps such as gallstones.

Calluses - thick, hardened areas of the skin, usually on the foot, caused by friction or pressure.

Cancellous tissue - the sponge-like tissue inside bones.

Cancer - abnormal cells which divide without control, which can invade nearby tissues or spread through the bloodstream and lymphatic system to other parts of the body.

Candidiasis - mild infection caused by the Candida fungus, which lives naturally in the gastrointestinal tract. Infection occurs when a change in the body, such as surgery, causes the fungus to overgrow suddenly.

Capillaries - tiny blood vessels between arteries and veins which distribute oxygen-rich blood to the body.

Capsule - the layer of cells around an organ, such as the capsule of the prostate. Also a formulation of pills.

Carcinogen - a substance that is known to cause cancer.

Cardiac - pertaining to the heart.

Cardiac arrest - the stopping of heartbeat.

Cardiac output - total amount of blood being pumped by the heart over a particular period of time.

Cardiology - the clinical study and practice of treating the heart.

Cardioversion - the procedure of applying electrical shock to the chest to change an abnormal heartbeat into a normal one.

Carotid arteries - the major arteries in the neck which supply blood to the brain.

Cartilage - a smooth material that covers bone ends of a joint to cushion the bone and allow the joint to move easily without pain.

Cathartics - laxatives.

Catheter - a flexible tube used to drain fluid from or inject fluid into the body. The most common catheter is the Foley catheter, used to drain urine from the bladder.

Cecostomy - tube which goes through the skin into the beginning of the large intestine to remove gas or feces; it is a short-term way to protect part of the colon while it heals after surgery.

Cecum - beginning of the large intestine; it is connected to the lower part of the small intestine, called the ileum.

Cellular pathology (also called cytopathology.) - the study of cellular alterations in disease.

Central auditory processing disorder - inability of individuals with normal hearing and intelligence to differentiate, recognize, or understand sounds.

Cerebellum - a large structure consisting of two halves (hemispheres) located in the lower part of the brain; responsible for the coordination of movement and balance.

Cerebral embolism - a blood clot from one part of the body that is carried by the bloodstream to the brain, where it blocks an artery.

Cerebral hemorrhage - bleeding within the brain.

Cerebral thrombosis - formation of a blood clot in an artery that supplies blood to the brain.

Cerebrovascular - pertaining to blood vessels in the brain.

Cerebrovascular accident (also called stroke) - an impeded blood supply to the brain.

Cerebrovascular occlusion - an obstruction in the blood vessel in the brain.

Cerebrum - consists of two parts (lobes), left and right, which form the largest and most developed part of the brain; initiation and coordination of all voluntary movement take place within the cerebrum. The basal ganglia are located immediately below the cerebrum.

Cervical intraepithelial neoplasia (CIN) - term used to classify the degree of precancerous change in cells of the cervix in a condition called cervical dysplasia.

Cesarean delivery (also called cesarean section or C-section) - surgical procedure to deliver a baby through an incision in the lower abdomen and uterus.

Chemosensory disorders - disorders or diseases of smell and/or taste.

Chlorhydria - too much hydrochloric acid in the stomach.

Cholangiography - series of x-rays of the bile ducts.

Cholecystitis - inflammation of the gallbladder wall.

Cholecystokinin - hormone released in the small intestine; causes muscles in the gallbladder and the colon to tighten and relax.

Cholelithiasis - a condition in which gallstones are present in the gallbladder.

Chorea - rapid, jerky, dance-like movement of the body.

Chyme - thick liquid made of partially digested food and stomach juices; made in the stomach and moves into the small intestine for further digestion.

Cineangiography - a procedure which involves taking moving pictures to show the passage of dye through blood vessels.

Circulatory system - pertaining to the heart and blood vessels, and the circulation of blood.

Claudication - pain or fatigue in arms and legs due to a poor supply of oxygen to the muscles.

Clinical trials - studies, involving patients, aimed at finding better ways to prevent, detect, diagnose, or treat cancer (disease).

Clostridium difficile (also called C. difficile) - bacteria naturally present in the large intestine which makes a substance that can cause a serious infection, called pseudomembranous colitis, in people taking antibiotics.

Coccydynia - pain around the coccyx.

Cochlea - snail-shaped structure in the inner ear which contains the organ of hearing.

Cochlear implant - medical device which bypasses damaged structures in the inner ear and directly stimulates auditory nerves to allow some deaf individuals to learn to hear and interpret sounds and speech.

Code blue or code red: A patient who requires CPR. (should insert CPR and explanation)

Cognitive dysfunction: refers to patients who have medical conditions that cloud their thinking.

Colectomy - partial or complete removal of the large bowel or colon.

Colon - large intestine.

Colon polyps - small, fleshy, mushroom-shaped growths in the colon.

Colonic inertia - condition of the colon when muscles do not work properly, causing constipation.

Colonoscopic polypectomy - removal of tumor-like growths (polyps) using a device inserted through a colonoscope.

Colony-stimulating factors - substances which stimulate the production of blood cells.

Coloproctectomy - proctocolectomy.

Colorectal cancer - cancer which occurs in the colon (large intestine) or the rectum (the end of the large intestine).

Colorectal transit study - test to see how food moves through the colon.

Colostrum - thin, white, first milk produced by the breasts during late pregnancy and for a few days after childbirth. It provides a nursing infant with essential nutrients and infection-fighting antibodies.

Common bile duct - tube which carries bile from the liver to the small intestine.

Common bile duct obstruction - blockage of the common bile duct, often caused by gallstones.

Compact tissue - the harder, outer tissue of bones.

Comparative pathology - the study of disease in animals and how it compares in humans.

Complementary medicine - any form of therapy used in combination with other alternative treatments or standard/conventional medicine.

Compliant: a cooperative person or patient; usually refers to medication usage

Composite resins (also called white fillings) - a tooth-colored plastic mixture filled with glass (silicon dioxide) which is used primarily for cosmetic improvements of the smile by changing the color of the teeth or reshaping disfigured teeth.

Computed tomography (CT or CAT scan) - a non-invasive procedure which takes cross-sectional images of the brain or other internal organs; to detect any abnormalities that may not show up on an ordinary x-ray.

Conductive hearing impairment - hearing loss caused by dysfunction of the outer or middle ear.

Cone biopsy (also called conization.) - a biopsy in which a larger cone-shaped piece of tissue is removed from the cervix by using the loop electrosurgical excision procedure or the cold knife cone biopsy procedure.

Constrict - tighten; narrow.

Continence - ability to hold in a bowel movement or urine.

Continent ileostomy - operation to create a pouch from part of the small intestine. Stool that collects in the pouch is removed by inserting a small tube through an opening made in the abdomen.

Contractions, labor - rhythmic tightening of the muscular wall of the uterus to push the fetus down through the vagina during childbirth.

Contraindications: a condition that makes a treatment unwise to proceed forward with.

Contusion - a bruise caused by a blow to the muscle, tendon, or ligament; caused when blood pools around the point of injury and discolors the skin.

Cornea - the clear, dome-shaped surface which covers the front of the eye.

Corneal curvature - the shape of the front surface of the eye.

Coronary artery bypass graft (CAB or CABG) - a surgical procedure in which a healthy blood vessel is transplanted from another part of the body into the heart to replace or bypass a diseased vessel.

Coronary heart disease - a condition in which the coronary arteries narrow from an accumulation of plaque (atherosclerosis) and cause a decrease in blood flow.

Coronary occlusion - an obstruction of one of the coronary arteries that decreases flow to the heart muscle.

Coronary thrombosis - the formation of a clot in one of the arteries that carry blood to the heart muscle.

Cortex - the outer layer of the cerebrum, densely packed with nerve cells.

Corticosteroids - potent anti-inflammatory hormones which are made naturally in the body or synthetically for use as drugs. The most commonly prescribed drug of this type is prednisone.

Craniectomy - excision of a part of the skull.

Craniotomy - surgical opening of the skull to gain access to the intracranial structures.

Crepitus - grinding noise or sensation within a joint.

Crown - a "cap" which covers a cracked or broken tooth, unfixed by a filling, to approximate its normal size and shape.

Cryoprostatectomy - freezing of the prostate through the use of liquid nitrogen probes guided by transrectal ultrasound of the prostate.

Cryothalamotomy - a surgical procedure in which a supercooled probe is inserted into a part of the brain called the thalamus in order to stop tremors.

Cryptosporidia - parasite which can cause gastrointestinal infection and diarrhea. (see also gastroenteritis)

Cued speech - method of communication which combines speech reading with a system of hand shapes placed near the mouth to help deaf or hard-of-hearing individuals differentiate words which look similar on the lips.

Cupping - the use of warmed glass jars to create suction on certain points of the body.

Cyanosis - bluish color in the skin due to insufficient oxygen levels in the blood.

Cyclic vomiting syndrome (CVS) - sudden, repeated attacks of severe vomiting (especially in children), nausea, and physical exhaustion with no apparent cause.

Cystic duct - tube which carries bile from the gallbladder into the common bile duct and the small intestine.

Cystic duct obstruction - blockage of the cystic duct, often caused by gallstones.

Cystometry - diagnostic procedure which measures bladder capacity and pressure changes as the bladder fills and empties.

Cystourethrocele - condition which results when the urethra and its supporting tissues weaken and drop into the vagina, leading to stress incontinence.

Cytology - the study of individual cells.

Cytomegalovirus (congenital) - one group of herpes viruses that infect humans and can cause a variety of clinical symptoms including deafness or hearing impairment; infection with the virus may be either before or after birth.

D/C - discharge.

DDS - Doctor of Dental Surgery.

Debridement - the surgical removal of foreign material and/or dead, damaged, or infected tissue from a wound or burn.

Decibel - unit which measures the intensity or loudness of sound.

Deciduous teeth - also known as "baby" or primary teeth.

Defecation - passage of bowel contents through the rectum and anus.

Defecography - x-ray of the anus and rectum to see how the muscles work to move stool.

Defibrillator - an electronic device used to establish normal heartbeat.

Degenerative disorder - a condition which causes body parts to deteriorate over time, such as muscular dystrophy, or arthritis, or aging.

Delayed gastric emptying - gastroparesis.

Delusions - a condition in which the patient has lost touch with reality and experiences hallucinations and misperceptions.

Dendrite - a thread-like extension from a nerve cell which serves as an antenna to receive messages from the axons of other nerve cells.

Dental amalgams (also called silver fillings) - a mixture of mercury (45 to 50 percent) and an alloy of silver, tin, and copper (50 to 55 percent) used to repair decayed teeth.

Dental fluorosis - a condition which results from drinking overly fluoridated water that often causes the teeth to become discolored and the enamel of the teeth to look spotted, pitted, or stained.

Dental implants - small dental appliances which are inserted into the upper and lower jaws to help restore a mouth which has little or no restorable teeth.

Dental pulp - the soft tissue around the tooth which contains nerves, blood vessels, and connective tissue.

Dental sealant - a thin plastic film that is painted on the chewing surfaces of the back teeth - the molars and premolars - to prevent tooth decay.

Depth perception - the ability to distinguish objects in a visual field.

Derm/derma - refers to skin

Dermatopathology - the study of the skin in diagnosing skin diseases.

Descending colon - the middle part of the colon located on the left side of the abdomen.

Diagnosis - identifying a disease by its signs, symptoms, and laboratory findings.

Diaphragm - primary muscle used for respiration, located just below the lung bases.

Diastolic blood pressure - the lowest blood pressure measure in the arteries, which occurs between heartbeats.

Diathermy machine - a piece of equipment used in the operating room to control bleeding.

Digestants - medicines which aid or stimulate digestion.

Digestion - process the body uses to break down food into simple substances for energy, growth, and cell repair.

Digestive system - the group of organs which break down foods into chemical components that the body can absorb and use for energy, and for building and repairing cells and tissues.

Digital rectal exam (DRE) - procedure in which the physician inserts a gloved finger into the rectum to examine the rectum and the prostate gland for signs of cancer.

Dilate - relax; expand.

Dilation and curettage (D & C) - a common gynecological surgery which consists of widening the cervical canal with a dilator and scraping the uterine cavity with a curette.

Diplopia - double vision.

Disc herniation (also called disc prolapse, disc bulge, or slipped disc) - disruption to the normal integrity of the intervertebral disc.

Distention - bloating or swelling of the abdomen.

Diuretic - a medication which lowers blood pressure.

Diverticulosis - condition which occurs when small pouches (diverticula) push outward through weak spots in the colon.

Diverticulum - small pouch in the colon. These pouches are not painful or harmful unless they become infected or irritated.

DMD - Doctor of Dental Medicine.

Dopa decarboxylase - an enzyme present in the body which converts levodopa to dopamine.

Dopamine - a chemical substance, a neurotransmitter, found in the brain which regulates movement, balance, and walking.

Dual energy x-ray absorptiometry (DEXA) - imaging technique which uses a very low dose of radiation to measure bone density for the diagnosis of osteoporosis.

Dumping syndrome (also called rapid gastric emptying) - condition which occurs when food moves too fast from the stomach into the small intestine.

Durable power of attorney - a legal document denoting a friend or family member as your legal guardian in case you are unable to make medical decisions for yourself.

Dysarthia - group of speech disorders caused by disturbances in the strength or coordination of the muscles of the speech mechanism as a result of damage to the brain or nerves.

Dysentery - infectious disease of the colon. Symptoms include bloody, mucus-filled diarrhea, abdominal pain, fever, and loss of fluids from the body.

Disequilibrium - any disturbance of balance.

Dysfluency - disruption in the smooth flow or expression of speech.

Dysgeusia - distortion or absence of the sense of taste.

Dyskinesia - an involuntary movement including athetosis and chorea.

Dysosmia - distortion or absence of the sense of smell.

Dysphagia - difficulty in swallowing.

Dysphonia - any impairment of the voice or difficulty speaking.

Dysplasia - an abnormality of growth.

Dyspnea - shortness of breath.

Dyspraxia of speech - partial loss of the ability to consistently pronounce words in individuals with normal muscle tone and coordination of the speech muscles.

Dysrhythmia - an abnormal heart rhythm.

Dystonia - a slow movement or extended spasm in a group of muscles.

Dystrophin - a protein found in normal muscle tissue.

Ear infection - presence and growth of bacteria or viruses in the ear.

Ear wax - yellow secretion from glands in the outer ear (cerumen) which keeps the skin of the ear dry and protected from infection. (explain cerumen)

Eczema - inflammation of the skin which causes itching and sometimes crusting, scaling, or blisters.

Ecto - outside or external;

Edema - swelling due to the build-up of fluid.

Ejection fraction - the measurement of the blood pumped out of the ventricle.

Elective surgery - an operation the patient chooses to have done, which is not required.

Electrocardiogram (ECG or EKG) - a test which records the electrical activity of the heart, shows abnormal rhythms (arrhythmias or dysrhythmias), and detects heart muscle damage.

Electrocoagulation - procedure which uses a high frequency current to harden tissue.

Electrodermal activity (EDA) - measures changes in perspiration rate.

Electrodessication - electrosurgery which destroys tissue.

Electrolytes - chemical substances in the body which conduct electricity and can be checked as clues to numerous problems.

Electromyogram (EMG) - a test to evaluate nerve and muscle function.

Electrophysiological study (EPS) - a cardiac catheterization to study electrical current in patients who have arrhythmias.

Electrosurgery - use of high-frequency electrical currents to destroy cancer cells.

Embolus/embolism - a "wandering" blood clot.

Emergency surgery - an operation performed immediately as a result of an urgent medical condition.

Encephalitis - a viral infection of the brain.

Endarterectomy - the surgical removal of plaque or blood clots in an artery.

Endo - when used as part of a word, means inward or internal.

Endocardium - the membrane which covers the inside surface of the heart.

Endocervical curettage (ECC) - a procedure which uses a narrow instrument called a curette to scrape the lining of the endocervical canal. This type of biopsy is usually completed along with the colposcopic biopsy.

Endodontist (also called a pulp specialist) - an Endodontist has undergone specialized training in performing root canal therapy.

Endolymph - fluid in the labyrinth - the organ of balance located in the inner ear.

Endometrial hyperplasia - abnormal thickening of the endometrium caused by excessive cell growth.

Endometrial implants - fragments of endometrium that relocate outside of the uterus, such as in the muscular wall of the uterus, ovaries, fallopian tubes, vagina, or intestine.

Endometrium - mucous membrane lining of the inner surface of the uterus which grows during each menstrual cycle and is shed in menstrual blood.

Endorphins - endogenous biomechanical substances implicated in the alleviation of pain, produced as a result of body stress.

Endoscopic papillotomy - see endoscopic sphincterotomy.

Endoscopic retrograde cholangiopancreatography (ERCP) - this procedure involves inserting an endoscope (viewing tube) through the stomach and into the small intestine. A special dye injected during this procedure shows the ducts in the biliary system.

Endoscopic sphincterotomy (also called endoscopic papillotomy) - operation to cut the muscle between the common bile duct and the pancreatic duct.

Endothelium - the layer of cells which lines the heart, blood vessels, and other body cavities.

Enema - liquid put into the rectum to clear out the bowel or to administer medications or food.

Enteral nutrition (also called tube feeding) - way to provide food through a tube placed in the nose, the stomach, or the small intestine. A tube in the nose is called a nasogastric or nasoenteral tube. A tube which goes through the skin into the stomach is called a gastrostomy or percutaneous endoscope.

Enterocele - hernia in the intestine.

Enteroscopy - examination of the small intestine with an endoscope.

Enterostomy - ostomy, or opening, into the intestine through the abdominal wall.

Entrapment neuropathy - pathological situation where a nerve is trapped in an abnormally-produced anatomical or physiological site.

Enzyme-linked immunosorbent assay (ELISA) - blood test used to find Helicobacter pylori bacteria; also used to diagnose an ulcer.

Eosinophilic gastroenteritis - infection and swelling of the lining of the stomach, small intestine, or large intestine.

Epicardium - the membrane which covers the outside of the heart.

Epidermis-the outer layer of skin

Epidural anesthetic - an anesthetic which is injected into the "epidural space" in the middle and lower back, just outside the spinal space, to numb the lower extremities.

Epinephrine - one of two chemicals (the other is norepinephrine) released by the adrenal gland which increase heart rate and force of the heart contraction, and narrows blood vessels.

Ergonomics - the study of how the workplace relates to human functions.

Eructation - belching.

Erythema nodosum - red swellings or sores on the lower legs during flare-ups of Crohn's disease and ulcerative colitis.

Erythrocyte sedimentation rate - blood test which measures the speed at which red blood cells settle on the bottom of a test tube; high sedimentation rate signals a possible inflammatory disease.

Esophageal reflux - see gastroesophageal reflux disease.

Esophageal ulcer - sore in the esophagus caused by long-term inflammation or damage from the residue of pills.

Esophageal varices - stretched veins in the esophagus which occur when the liver is not working properly.

Estrogen - a hormone secreted by the ovaries which affects many aspects of the female body, including menstrual cycles and pregnancy.

Estrogen replacement therapy (ERT) - use of the female hormone estrogen to replace that which the body no longer produces naturally after medical or surgical menopause.

Euphoria - a feeling of well-being or elation; may be drug related.

Excisional - cutting away cancerous tissue with a scalpel, or other instruments, to completely remove it and possibly some surrounding tissue. There are many types of excisional surgeries, each named for the particular area of the body in which they are performed.

Excrete - to get rid of waste from the body.

Expectant management (also called expectant therapy) - "watchful waiting" or close monitoring of a disease by a physician instead of immediate treatment.

Expiration - exhaling; giving off carbon dioxide.

Extensor muscle - any muscle which causes the straightening of a limb or other part.

External urethral sphincter muscle - a voluntary and involuntary ring-like band of muscle fibers which voluntarily contract to stop urinating.

Extrahepatic biliary tree - bile ducts located outside the liver.

Extrapyramidal system - system consisting of nerve cells, nerve tracts, and pathways which connect the cerebral cortex, basal ganglia, thalamus, cerebellum, reticular formation, and spinal neurons that is concerned with the regulation of reflex movements such as balance and walking.

Extrinsic asthma - asthma which is triggered by an allergic reaction - usually to something that is inhaled.

Fallopian tubes - two thin tubes which extend from each side of the uterus, toward the ovaries, as a passageway for eggs and sperm.

False negative report - a negative result when in reality it is positive in nature.

False positive report - a positive result when in reality it is negative in nature.

Fascinoma - something doctors don't see often; a fascinating case

Fatty liver (also called steatosis) - build-up of fat in liver cells.

Fecal incontinence - being unable to hold stool in the colon and rectum.

Fecal occult blood test - test to check for hidden blood in stool; screening test for cancer of the colon or rectum.

Femur - thighbone.

Festination - walking with a series of quick, small, shuffling steps as if hurrying forward to keep balance.

Fibrillation - rapid contractions of the heart muscle.

Fibroadenoma - noncancerous, firm, rubbery lump in the breast which is painless and moves around easily when touched.

Fibrocystic breasts - noncancerous condition in which small lumps and cysts develop in the breasts.

Fibroids - noncancerous growths in, on, or within the walls of the uterus.

Fibrosis - the growth of scar tissue possibly due to infection, inflammation, injury, or even healing.

Fibrositis - a benign, intermittently recurring, and protracted disease process, with a lack of underlying pathology.

Fine needle aspiration - the use of a thin, hollow needle to withdraw tissue from the body.

Flexor muscle - any muscle which causes the bending of a limb or other body part.

Flutter - ineffective contractions of the heart muscles.

Follicle-stimulating hormone (FSH) - hormone secreted by the pituitary gland in the brain which stimulates the growth and maturation of eggs in females and sperm in males, and sex hormone production in both males and females.

Food exchanges - a way to help people stay on special food plans by letting them replace items from one food group with items from another group.

Food intolerance - an adverse food-induced reaction that does not involve the immune system.

Free skin graft - the detaching of healthy skin from one part of the body to repair areas of lost or damaged skin in another part of the body.

Frozen shoulder - alternate name for "capsulitis" of the shoulder, which undergoes four distinct stages: pain; pain and stiffness; stiffness; resolution. explain capsulitis

Functional disorders (also called motility disorders) - conditions which result from poor nerve and muscle function.

Functional incontinence - difficulty reaching a restroom in time because of physical conditions such as arthritis.

Gait - pattern of locomotion.

Gamete intrafallopian transfer (GIFT) - method of treating infertility by removing eggs from a woman's ovaries, combining them with sperm from her partner or a donor in the laboratory, and placing the eggs and sperm together in one of her fallopian tubes, where fertilization can occur.

Ganglion - a cluster of nerve cells.

Ganglion cysts - non-cancerous, fluid-filled cysts are common masses or lumps in the hand and usually found on the back of the wrist.

Gardner's syndrome - condition in which many polyps form throughout the digestive tract.

Gas - air which comes from the normal breakdown of food and is passed out of the body through the rectum (flatus) or the mouth (belch).

Gastric - related to the stomach.

Gastric juices - liquids produced in the stomach to help break down food and kill bacteria.

Gastrocolic reflex - increase of muscle movement in the gastrointestinal tract when food enters an empty stomach; may cause the urge to have a bowel movement right after eating.

Gastroenteritis - infection or irritation of the stomach and intestines, which may be caused by bacteria or parasites from spoiled food or unclean water, or eating food which irritates the stomach lining, and emotional upsets such as anger, fear, or stress.

Gastroenterologist - physician who specializes in digestive diseases.

Gastroenterology - field of medicine concerned with the function and disorders of the digestive system.

Gastrointestinal (GI) tract (also called the alimentary canal or digestive tract) - a large, muscular tube which extends from the mouth to the anus, where the movement of muscles and release of hormones and enzymes digest food.

Gastrostomy - an artificial opening from the stomach to a hole (stoma) in the abdomen where a feeding tube is inserted.

Gated blood pool scan - a nuclear scan to see how the heart wall moves and how much blood is expelled with each heart beat just after the patient has walked on a treadmill or ridden on a stationary bike.

General anesthetic - an anesthetic which causes the patient to become unconscious during surgery.

Genes - basic, functional units of heredity, each occupying a specific place on a chromosome. explain chromosome

Genital herpes - a sexually transmitted disease caused by the herpes simplex virus.
Genu valgum - commonly known as "knock knees."

Genu varum - commonly known as "bowed legs."

GI series/tract: tests referring to the digestive system

Glucose - a simple sugar, which is the body's main source of energy.

Gluten intolerance - see Celiac disease.

Gluten sensitive enteropathy (also called celiac sprue or celiac disease) - a sensitivity to gluten, a wheat protein. Individuals with this disease must avoid gluten-containing grains, which include all forms of wheat, oats, barley, and rye.

Gluteus maximus - the large, superficial buttock muscle.

Glycogen - converted glucose for storage. Glycogen plays a role in controlling blood sugar levels.

Gonadotropins - luteinizing hormone and follicle-stimulating hormone, produced by the pituitary gland.

Gonads - ovaries and testes.

Grades of movement - standardized means of documenting techniques of mobilization, relating it to the true feel of joint movement.

Grading - a process for classifying cancer cells to determine the growth rate of the tumor. The cancer cells are measured by how closely they look like normal cells.

Granuloma - mass of red, irritated tissue in the gastrointestinal tract found in Crohn's disease.

Granulomatous colitis - another name for Crohn's disease of the colon.

Granulomatous enteritis - another name for Crohn's disease of the small intestine.

Gray matter - the darker-colored tissues of the central nervous system; in the brain, the gray matter includes the cerebral cortex, the thalamus, the basal ganglia, and the outer layers of the cerebellum.

Guided imagery - envisioning a certain goal to help cope with health problems.

Gustation - act or sensation of tasting.

H2-blockers - medicines which reduce the amount of acid the stomach produces by blocking histamine2, which signals the stomach to make acid.

Hair cells - sensory cells of the inner ear, which are topped with hair-like structures (stereocilia), which transform the mechanical energy of sound waves into nerve impulses.

Halitosis - an oral health condition characterized by consistently odorous breath.

Hallux rigidus - loss of range of the movement of the metatarsophalangeal (MTP) joint of the first (great) toe.

Hamstrings - muscles located in the posterior compartment of the thigh.

Hay fever - see rhinitis.

Headache, primary - includes tension (muscular contraction), vascular (migraine), and cluster headaches not caused by other underlying medical conditions.

Headache, secondary (also called traction headaches or inflammatory headaches) - includes headaches which result from other medical conditions.

Hearing - series of events in which sound waves in the air are converted to electrical signals which are sent as nerve impulses to the brain, where they are interpreted.

Hearing aid - electronic device which brings amplified sound to the ear.

Hearing disorder - disruption in the normal hearing process; sound waves are not converted to electrical signals and nerve impulses are not transmitted to the brain to be interpreted.

Heart block - interrupted electrical impulse to heart muscles.

Heart valve prolapse - a condition of the heart valve in which it is partially open when it should be closed.

Heartbeat - one complete contraction of the heart.

Heart-lung machine - a machine which performs for the heart during open heart surgery.

Heberden's nodes - bony swellings around the margins of joints, associated with degenerative changes of arthritis.

Hema/hemo - refers to "blood" when used as a part of a word.

Hematopathology - the study of blood, bone marrow and the organs and tissues which use blood cells to perform their functions.

Hemorrhage - the medical term for bleeding.

Hemorrhoidectomy - the surgical removal of hemorrhoids.

Hepatitis E - virus spread mostly through unclean water. This type of hepatitis is common in developing countries and has not occurred in the United States.

Hepatologist - physician who specializes in liver diseases.

Hepatology - field of medicine concerned with the functions and disorders of the liver.

High density lipoprotein (HDL) - the "good" cholesterol which promotes breakdown and removal of cholesterol from the body.

Hip - the region on each side of the pelvis which is made up of three sections: ilium, ischium, and pubis; the upper part of the femur (upper leg bone) fits into the hip via a ball-and-socket joint; the socket is a cup-shaped bone of the pelvis called the acetab.

Histamine - a chemical present in cells throughout the body that is released during an allergic reaction.

Hormone replacement therapy (HRT) - use of the female hormones estrogen and progestin (a synthetic form of progesterone) to replace those the body no longer produces after menopause.

Hormone therapy - the use of hormones, medications, or surgery to suppress (block) or mimic hormones and alter the growth of hormone-sensitive cancer.

Hormones - chemical substances created by the body which control numerous body functions.

Human chorionic gonadotropin - hormone produced by the placenta during early pregnancy.

Human papillomaviruses (HPV's) - a group of viruses that can cause warts. Some HPV's are sexually transmitted and cause wart-like growths on the genitals. HPV is a major risk factor for cervical cancer.

Hyaloid canal - narrow passageway which allows blood to flow through the eye.

Hydrocortisone - a hormone secreted by the adrenal cortex which affects metabolism.

Hydrogen breath test - test for lactose intolerance that measures breath samples for too much hydrogen.

Hydrotherapy - use of water as a medical treatment

Hyperextension - active or passive force which takes the joint into extension, but beyond its normal physiological range.

Hyperglycemia - high levels of glucose in the blood.

Hyperopia - farsightedness

Hyperplasia, endometrial - excessive growth of cells in the endometrium.

Hypertension - abnormally high blood pressure.

Hypertrophy - an increase in the size of tissue.

Hypo - when used as part of a word means under or below.

Hypogeusia - diminished sensitivity to taste.

add Hypoglycemia

Hypomobility - a decrease in the normal range of joint movement.

Hyposmia - diminished sensitivity to smell.

Hypotension - abnormally low blood pressure.

Hypothalamus - small structure at the base of the brain which regulates many body functions, including appetite and body temperature.

Hypoxia - depletion of oxygen in the cells and tissues.

Hysteroscope - visual examination of the canal of the cervix and the interior of the uterus using a viewing instrument (hysteroscope) inserted through the vagina.

Ice therapy - cooling of deeper tissues.

Idiopathic - of unknown origin.

Indication - a condition which requires a specific treatment or procedure.

Ileal - related to the ileum, the lowest end of the small intestine.

Ileocolitis - irritation of the lower part of the small intestine (ileum) and colon.

Ileostomy - operation that makes it possible for stool to leave the body, after the colon and rectum are removed, in which an opening is made in the abdomen and the bottom of the small intestine (ileum) attaches to it.

Ileum - lower end of the small intestine.

In vitro - artificial environment.

Intravenous (IV) - inserted into your vein.

Imaging - tests or evaluation procedures which produce pictures of areas inside the body.

Immobilization - preventing movement to allow for natural healing to take place.

Immune system - complex network of specialized cells and organs which work together to defend the body against attacks by "foreign" invaders such as bacteria and viruses.

Immunoglobulin E (IgE) - a type of antibody, formed to protect the body from infection, which attaches to mast cells in the respiratory and intestinal tracts and may cause allergic rhinitis, asthma, or eczema.

Immunoglobulins - antibodies or proteins found in blood and tissue fluids produced by cells of the immune system to bind to substances in the body which are recognized as foreign antigens. Immunoglobulins sometimes bind to antigens that are not necessarily a threat to health.

Immunology - the study of the body's immune system, its functions and disorders.

Immunosuppressive medications - medications which suppress the body's immune system; often used to minimize rejection of transplanted organs.

Immunotherapy - treatment of allergy to substances such as pollens, house dust mites, fungi, and stinging insect venom by giving gradually increasing doses of the substance, or allergen, to which the person is allergic.

Immunotherapy (also called biological therapy) - treatment which uses the body's natural defenses to fight disease.

Impaction - trapping of an object in a body passage, such as stones in the bile duct or hardened stool in the colon.

Impaired glucose tolerance (IGT) - a condition in which blood sugar levels are higher than normal, but are not high enough to be classified as diabetes; a risk factor for type 2 diabetes.

Impedance plethysmography - a test to evaluate blood flow through the leg.

In vitro fertilization - treatment for infertility in which a woman's egg is fertilized, outside her body, with her partner's sperm or sperm from a donor.

Infection - the invasion of the body by microorganisms that cause disease.

Inferior vena cava - the large blood vessel (vein) which returns blood from the legs and abdomen to the heart.

Inflammation - characteristic reaction of tissues to injury or disease marked by four signs: swelling, redness, heat, and pain.

Inflammatory bowel disease (IBD) - long-lasting problem which causes irritation and ulcers in the gastrointestinal (GI) tract; the most common disorders are ulcerative colitis and Crohn's disease.

Informed consent form - a form signed by the patient prior to surgery which explains everything involved in the surgery, including its risks.

Inner ear - part of the ear which contains both the organ of hearing (the cochlea) and the organ of balance (the labyrinth).

Inotropic medications - medications which increase strength of the contractions in the heart.

Inpatient surgery - surgery which requires the patient to be admitted for a stay in the hospital.

Inspiration - inhaling; taking in oxygen.

Insulin - a hormone produced by the pancreas. Insulin affects the amount of glucose absorbed by the liver.

Insulin-dependent diabetes - see type 1 diabetes.

Insulin-resistance - partial blocking of the effect of insulin.

Intercostal muscles - muscles lying between ribs; often injured by rotary stress of the thorax.

Interferon - a biological response modifier which stimulates the growth of certain disease-fighting blood cells in the immune system; used to treat cancer and other diseases.

Interleukin-2 - a biological response modifier which stimulates the growth of certain blood cells in the immune system that can fight cancer and other diseases.

Internal derangement of the joint - a dislocated jaw or displaced disc, or injury to the condyle (round edges of the jaw).

Intervertebral disc - disc which forms a cartilaginous joint between the vertebral bodies to provide shock absorption.

Intestinal flora - bacteria, yeasts, and fungi which grow normally in the intestines.

Intestinal mucosa - surface lining of the intestines where the cells absorb nutrients.

Intestines - see large intestine and small intestine.

Intolerance - allergy or sensitivity to a food, drug, or other substance.

Intra-articular - within the joint.

Intrauterine insemination - treatment for infertility in which semen is introduced into the uterus via a slim tube inserted through the vagina.

Intravenous line (IV) - a thin plastic tube inserted into a vein (usually in the patient's forearm) through which a volume of fluid is injected into the bloodstream.

Intrinsic asthma - asthma that has no apparent external cause.

Invasive cancer - cancer which begins in one area and then spreads deeper into the tissues of that area.

Investigational new drug - a drug allowed by the U.S. Food and Drug Administration (FDA) to be used in clinical trials, but not approved for sale to the general public.

Ischemia - decreased flow of oxygenated blood to an organ due to obstruction in an artery.

Ischemic heart disease - coronary artery disease or coronary heart disease caused by narrowing of the coronary arteries and decreased blood flow to the heart.

Isometric - muscle contraction without movement at the joint.

Isthmus - tissue which connects the two lobes of the thyroid.

-itis - refers to inflammation when used as a part of a word.

Jejunum - middle section of the small intestine between the duodenum and ileum.

Joint - where the ends of two or more bones meet.

Joint locking - extremely painful condition usually caused by entrapment of a loose body within the joint, between the joint surfaces.

Jordan frame - specialized stretcher developed for transport of patients with a suspected spinal injury.

Jugular veins - veins which carry blood from the head back to the heart.

Jumper's knee - associated with a small area of degeneration of the tendon attachment at the lower pole of the patella (knee-cap), characterized by pain and extreme local tenderness.

Keratitis - inflammation of the cornea.

Ketoacidosis - high blood glucose, often caused by illness or too little insulin.

Ketone - break down of production of fat which accumulates in the blood as a result of inadequate insulin or inadequate calorie intake.

Knee reconstruction - surgical restoration of biomechanics of the knee.

Labyrinth - organ of balance located in the inner ear; consists of three semicircular canals and the vestibule.

Labyrinthine hydrops - excessive fluid in the organ of balance (labyrinth) which can cause pressure or fullness in the ears, hearing loss, dizziness, and loss of balance.

Lactase - enzyme in the small intestine needed to digest milk sugar (lactose).

Lactase deficiency - lack of the lactase enzyme.

Lactose - sugar found in milk, which the body breaks down into galactose and glucose.

Laminectomy - surgical procedure which includes removal of a portion of the lamina (a plate or layer of the bone), to provide more room in the vertebral canal; usually performed to treat disc herniation or spinal canal stenosis.

Landau-Kleffner syndrome - a childhood disorder of unknown origin which can be identified by gradual or sudden loss of the ability to understand and use spoken language.

Language - system for communicating ideas and feelings using sounds, gestures, signs, or marks.

Laparoscope - thin tube with a tiny video camera attached, used to look inside the body and see the surface of organs.

Laparoscopic cholecystectomy - an operation to remove the gallbladder. The doctor inserts a laparoscope, and other surgical instruments, through small holes in the abdomen. The camera allows the doctor to see the gallbladder on a television screen.

Laparoscopic lymph node sampling - lymph nodes are removed through a viewing tube called a laparoscope, which is inserted through a small incision in the abdomen.

Laparoscopic lymphadenectomy - the removal of pelvic lymph nodes with a laparoscope through small incisions in the lower abdominal region.

Laparotomy - a surgical procedure which involves an incision into the abdominal cavity; often used when making a diagnosis when less invasive tests are difficult or not possible.

Large intestine - part of the intestine that goes from the cecum to the rectum.

Laryngeal neoplasms - abnormal growths in the larynx (voice box) which can be cancerous or noncancerous.

Laryngeal nodules - noncancerous, callous-like growths on the inner parts of the vocal folds.

Laryngeal paralysis - loss of function or feeling of one or both of the vocal folds.

Laryngectomy - surgery to remove part or all of the larynx or voice box.

Larynx - valve structure between the trachea (windpipe) and the pharynx (the upper throat); the primary organ of voice production.

Lavage - the process of washing (cleaning) of an organ such as the bowel or stomach.

Laxatives (also called cathartics) - medications to relieve long-term constipation.

Lens (also called crystalline lens) - the transparent structure inside the eye that focuses light rays onto the retina.

Lesion - an injury or wound.

Levodopa (L-dopa) - an effective anti-Parkinson drug which is changed into dopamine in the brain.

Lewy body - a pink-staining sphere, found in the bodies of dying cells, which is considered to be a marker for Parkinson's disease.

Ligament - a flexible band of fibrous tissue that connects the bones and bind the joints together.

Lipid - a fatty substance in the blood.
Lipo - when used as part of a word, means fat.

Lipoproteins - transporters of fatty substances in the blood.

Lithotripsy, extracorporeal shock wave (ESWL) - method of breaking up bile stones and gallstones with a specialized tool and shock waves.

Liver - largest organ in the body; performs important functions such as making bile, changing food into energy, and cleaning alcohol and poisons from the blood.

Liver enzyme tests (also called liver function tests [LFT's]) - blood tests which indicate the functioning of the liver and biliary system.

Living will - a legal document which states your medical preferences for treatment and resuscitation in the event you can no longer speak or act for yourself.

Lobectomy - removal of a lobe of the lung for cancer, benign tumors, or infections.

Lobule - a subdivision of a lobe or a small lobe.

Local anesthesia (a.k.a. Local) - anesthetic medicine injected into the site of the operation to temporarily numb that area.

Loop electrosurgical excision procedure (LEEP) - procedure for treating cervical dysplasia, in which a fine wire loop and low-energy current are used to remove abnormal tissue from the cervix.

Loose body - name given to an object located within a joint which has become detached.

Low blood glucose - a condition which occurs in people with Diabetes when their blood glucose levels are too low.

Low density lipoprotein (LDL) - a blood substance with large amounts of cholesterol.

Lower (gastrointestinal) GI series (also called barium enema x-ray, LGI,) - x-rays of the rectum, colon, and lower part of the small intestine.

Lower back (also called lumbar spine) - a complex structure which connects the upper body to the lower body; consists of vertebrae, disks, spinal cord, and nerves.

Lower esophageal sphincter - muscle between the esophagus and stomach.

Lumen - the hollow area within a tube.

Lumpectomy - a surgical procedure in which only the tumor and a small area of surrounding tissue are removed.

Lung volume - the amount of air the lungs hold.

Luteinizing hormone (LH) - hormone secreted by the pituitary gland in the brain which stimulates the growth and maturation of eggs in females and sperm in males.

Lymph nodes (also called lymph glands.) - small glands clustered in the neck, armpits, abdomen, and groin, which supply infection-fighting cells to the bloodstream, and filter out bacteria and other antigens.

Lymphadenectomy - a procedure in which lymph nodes are taken from the body for purposes of diagnosing or staging cancer.

Lymphatic system - tissues and organs, including bone marrow, spleen, thymus, and lymph nodes, which produce, store, and carry white blood cells to fight infection and disease.

Lymphocyte - any one of a group of white blood cells of crucial importance to the adaptive part of the body's immune system.

Magnetic field therapy - using an alternating magnetic field to generate an electric current inside the tissues, resulting in changes to blood flow.

Magnetic resonance imaging (MRI) - a non-invasive procedure which produces a two-dimensional view of an internal organ or structure, especially the brain and spinal cord.

Malabsorption syndromes - conditions which happen when the small intestine cannot absorb nutrients from foods.

Malignant tumor - a mass of cancer cells which may invade surrounding tissues or spread to distant areas of the body.

Mallet finger - common condition due to the rupture of the long extensor tendon of the finger.

Malocclusion - an orthodontic problem which means "bad bite," including crowded, missing, or crooked teeth, extra teeth, or a misaligned jaw.

Mammogram - a low-dose x-ray of the breast.

Manometry - tests which measure muscle pressure and movements in the gastrointestinal (GI) tract.

Massage - mechanical form of therapy in which the soft tissues are made more pliable with different techniques, promoting increased blood flow and subsequent healing.

Mast cells - cells which synthesize and store histamines, found in most body tissues, particularly just below the epithelial surfaces, serous cavities, and around blood vessels.

Mastitis - infection of the milk ducts in the breast.

Mastoid - back portion of the temporal bone behind the ear.

Mastoid surgery - surgical procedure to remove an infection from the mastoid bone.

McMurray test - test for assessing the knee for the presence of a medial meniscal lesion.

Meal plan - a guide to help people get the proper amount of calories, carbohydrates, proteins, fats, vitamins, minerals, and fiber in their diet.

Mean blood pressure - average blood pressure, taking account of the rise and fall that occurs with each heartbeat. It is often estimated by multiplying the diastolic pressure by two, adding the systolic pressure, and then dividing this sum by three.

Median nerve - large nerve, comprising segments from the cervical spine, which is involved in neural function of the upper limb; commonly entrapped in the carpal tunnel of the wrist, to create carpal tunnel syndrome. (reference Carpal Tunnel Syndrome)

Megacolon - huge, swollen colon; results from severe constipation.

Meige syndrome - a movement disorder which can involve excessive eye blinking (blepharospasm) with involuntary movements of the jaw muscles, lips, and tongue.

Melanocytes - cells in the skin which produce and contain the pigment called melanin.

Melena - blood in the stool.

Membrane - a thin layer of tissue which covers or lines part of your body.

Ménétrier's disease - long-term disorder which causes large coiled folds in the stomach.

Meniscus - a part of the cartilage in the knees and other joints.

Metastasis - when cancer cells spread beyond the site where the cancer originated.

Micrographia - a change in handwriting with the script becoming smaller and more cramped.

Middle ear - part of the ear which includes the eardrum and three tiny bones of the middle ear, ending at the round window that leads to the inner ear.

Minimally invasive surgery - any technique involving surgery that does not require a large incision.

Minor surgery: usually refers to surgery without general anesthesia.

Miosis - constriction of the pupil.

Misarticulation - inaccurately produced speech sound (phoneme) or sounds.

Miscarriage - spontaneous termination of a pregnancy before the fetus has developed enough to survive outside the uterus.

Mitral valve - the valve which controls blood flow between the left atrium and left ventricle in the heart.

Modified radical mastectomy - the removal of the breast, some lymph nodes in the armpit, and sometimes part of the chest wall muscles.

Monoamine oxidase (MAO) - an enzyme which makes nervous system hormones inactive.

Monoclonal antibodies - substances which can locate and bind to cancer cells wherever they are in the body.

Monounsaturated fats - dietary fats, such as olive oil or canola oil, which do not seem to have any affect on blood cholesterol.

Morton's neuroma - painful condition of one of the digital branches of the plantar nerves in the foot.

Motility - movement of food through the digestive tract.

Motion sickness - dizziness, sweating, nausea, vomiting, and generalized discomfort experienced when an individual is in motion.

Motor speech disorders - group of disorders caused by the inability to accurately produce speech sounds (phonemes).

Moxibustion - the burning of herbal leaves on or near the body.

Mucosal lining - lining of gastrointestinal (GI) tract organs that makes mucus.

Mucosal protective drugs - medicines which protect the stomach lining from acid.

Mucous colitis - see irritable bowel syndrome (IBS).

Mucus - clear liquid made by the intestines which coats and protects tissues in the gastrointestinal tract.

Multifidus - deep lumbar spine muscle which stabilizes the lumbar spine.

Murmur - a blowing or rasping sound heard while listening to the heart; may or may not indicate problems within the heart or circulatory system.

Musculoskeletal system - the complex system which includes the bones, joints, ligaments, tendons, muscles, and nerves.

Mycoplasma - very common sexually transmitted disease or urinary tract infection caused by a bacteria-like organism in the urethra and reproductive system.

Mydriasis - dilation of the pupil.

Myelogram - a specific x-ray study which uses an injection of a dye or contrast material into the spinal canal to allow careful evaluation of the spinal canal and nerve roots.

Myocardial infarction (also called heart attack) - occurs when one of more regions of the heart muscle experience a severe or prolonged decrease in oxygen supply caused by a blocked blood flow to the heart muscle.

Myocardial ischemia - insufficient blood flow to part of the heart.

Myocardium - the muscle wall of the heart.

Myoclonus - jerking, involuntary movements of the arms and legs; may occur normally during sleep.

Myofascial pain - the most common form of temporomandibular disorder (TMD); discomfort or pain in the muscles that control jaw function and the neck and shoulder muscles.

Myofascial trigger point - areas of focal muscle tenderness and spasm.

Myomectomy - surgical procedure done to remove fibroids from the uterus while leaving the uterus intact.

Myopia - nearsightedness.

National Cancer Institute - the U.S. government agency for cancer research and information.

Near point of accommodation - the closest point in front of the eyes at which an object may be clearly focused.

Near point of convergence - the maximum extent the two eyes can be turned inward.

Needle aspiration (of the breast) - uses a thin needle and syringe to collect tissue or drain a lump after using a local anesthetic.

Needle biopsy - biopsy procedure in which a small sample of tissue is removed through a hollow needle.

Neoadjuvant therapy - chemotherapy given before surgery or radiation therapy.

Nephrectomy - surgical removal of the kidney.

Nephropathy - diabetic kidney disease.

Nerve conduction test - procedure to determine nerve impulse generation.

Nerve sparing technique - a surgical technique often used during a radial prostatectomy in which one or both of the neurovascular bundles controlling erections are spared.

Neural mobilization - techniques by which neural tissues are moved, either by movement relative to their surroundings or by tension development.

Neural plasticity - ability of the brain and/or certain parts of the nervous system to change in order to adapt to new conditions, such as an injury.

Neural prostheses - devices which substitute for an injured or diseased part of the nervous system to enhance the function.

Neural stimulation - to activate or energize a nerve through an external source.

Neural tube defect - type of birth defect, such as spina bifida, which results from failure of the spinal cord or brain to develop normally in a fetus.

Neuralgia - a painful condition caused by disorders of the nervous system.

Neuritis - inflammation of a nerve or nerves.

Neurofibromatosis (Von Recklinghausen's) - a group of inherited disorders in which noncancerous tumors grow on several nerves, which may include the hearing nerve.

Neurogenic - of nerve origin.

Neurogenic communication disorder - inability to exchange information with others because of hearing, speech, and/or language problems caused by impairment of the nervous system.

Neuron - a cell specialized to conduct and generate electrical impulses and to carry information from one part of the brain to another.

Neuropathology - the study of the nervous system.

Neurotransmitters - chemical substances which carry impulses from one nerve cell to another; found in the space (synapse) that separates the transmitting neuron's terminal (axon) from the receiving neuron's terminal (dendrite).

Nissen fundoplication - operation to sew the top of the stomach (fundus) around the esophagus; used to stop stomach contents from flowing back into the esophagus (reflux) and to repair a hiatal hernia.

Noise-induced hearing loss - hearing loss that is caused either by a one-time or repeated exposure to very loud sound(s).

Non-insulin dependent diabetes - see type 2 diabetes.

Noninvasive procedures - a diagnostic effort or treatment which does not require entering the body or puncturing the skin.

Non-steroidal anti-inflammatory drugs (NsAIDs) - medications which produce antipyretic, analgesic, and anti-inflammatory effects.

Nonsyndromic hereditary hearing impairment - hearing loss or deafness that is inherited and is not associated with other inherited clinical characteristics.

Nonulcer dyspepsia - constant pain or discomfort in the upper gastrointestinal tract.

Norepinephrine - a neurotransmitter found mainly in areas of the brain which are involved in governing autonomic nervous system activity, especially blood pressure and heart rate.

Norwalk virus - virus which may cause gastrointestinal (GI) infection and diarrhea.

Nuclear medicine - a specialized area of radiology that uses very small amounts of radioactive substances to examine organ function and structure.

Nutrients - proteins, carbohydrates, fats, vitamins, and minerals provided by food and necessary for growth and the maintenance of life.

Obstruction - blockage in the gastrointestinal tract which prevents the flow of liquids or solids.

Occluded artery - an artery which is narrowed by plaque that impedes blood flow.

Occult - disease or symptoms which are not readily detectable by physical examination or laboratory tests.

Occult bleeding - blood in stool which is not visible to the naked eye.

Ocular hypertension - high (greater than 21 mm Hg) intraocular pressure.

Odorant - substance that stimulates the sense of smell.

Olfaction - the act of smelling.

Olfactometer - device for estimating the intensity of the sense of smell.

Oncogenes - genes which promote normal cell division.

Oncologist - a physician who specializes in treating cancer, including surgical oncologist, radiation oncologist, pediatric oncologist, gynecologic oncologist, and medical oncologist.

On-off effect (also called on-off phenomena) - a change in the patient's condition, with sometimes rapid fluctuations between uncontrolled movements and normal movement, usually occurring after long-term use of Levodopa and probably caused by changes in the ability to respond to this drug.

Oophorectomy - surgery to remove one or both ovaries.

Open surgery - cutting the skin and tissues during surgery to expose a full view of the structures and organs involved in the procedure.

Open-set speech recognition - understanding speech without visual clues.

Optional surgery - an operation the patient chooses to have done, which may not be essential to the continuation or quality of life.

Oral and maxillofacial surgeon - orthopedic facial surgeon who is responsible for treating a wide variety of dental problems, including the removal of impacted teeth and reconstructive facial surgery.

Oral dissolution therapy - method of dissolving cholesterol gallstones.

Orchiectomy (also called castration) - the surgical removal of the testicles.

Oropharynx - the part of the throat at the back of the mouth.

Orthodontics - orthodontics is the dental specialty which focuses on the development, prevention, and correction of irregularities of the teeth, bite, and jaw.

Orthokeratology - the use of contact lenses to change the shape of the cornea in order to correct refractive error.

Orthopedic surgeon (also called orthopedist) - the physician who diagnoses, treats, and manages the rehabilitation process, and provides prevention protocols for patients who suffer from injury or disease in any of the components of the musculoskeletal system.

Orthopedic surgery (also called orthopedics) - the medical specialty devoted to the diagnosis, treatment, rehabilitation, and prevention of injuries and diseases of the body's musculoskeletal system.

Orthostatic hypotension - a large decrease in blood pressure upon standing; may result in fainting.

Osteitis pubis - inflammatory reaction in the pubic symphysis, usually as a result of overuse; pain is felt in the groin region, thigh, lower abdomen, and around the perineal structures.

Osteoblast - cell found in bone; its function is to form the tissue and minerals which give bone its strength.

Osteoclast - very large cell formed in bone marrow; its function is to absorb and remove unwanted tissue.

Osteocyte - cell found within the bone; its function is to help maintain bone as living tissue.

Osteophyte - an outgrowth of bone.

Otoacoustic emissions - low-intensity sounds produced by the inner ear which can be quickly measured with a sensitive microphone placed in the ear canal.

Otolaryngologist - physician/surgeon who specializes in diseases of the ears, nose, throat, head, and neck.

Otologist - physician/surgeon who specializes in diseases of the ear.

Otosclerosis - abnormal growth of bone in the inner ear, which prevents structures within the ear from working properly, resulting in a gradual loss of hearing.

Ototoxic drugs - drugs which can damage the hearing and balance organs located in the inner ear.

Outer ear - external portion of the ear, consisting of the pinna, or auricle, and the ear canal.

Outpatient surgery - surgery which allows the patient to go home the same day.

Ovaries - pair of small glands, located on either side of the uterus, in which egg cells develop and are stored and thc female sex hormones estrogen and progesterone are produced.

Overflow incontinence - the leakage of small amounts of urine from a bladder that is always full.

Ovulation - release of a mature egg from an ovary.

Oxytocin - hormone produced by the pituitary gland which stimulates contractions of the uterus during labor and release of milk during breastfeeding.

Pacemaker - an electronic device which is surgically implanted into the patient's heart and chest to regulate heartbeat.

Pain threshold - the point at which a stimulus causes a feeling of pain.

Palliative treatment - therapy which relieves symptoms, such as pain, but does not alter the course of the disease; its primary purpose is to improvc the quality of life.

Pallidotomy - a surgical procedure in which a part of the brain, called the globus pallidus, is lesioned in order to improve symptoms of tremor, rigidity, and bradykinesia.

Palming - an imaging technique involving the visualization of color.

Palpitation - sensation of rapid heartbeats.

Palsy - paralysis of a muscle or group of muscles.

Pancreas - long gland which lies behind the stomach; manufactures insulin and digestive enzymes.

Papillary stenosis - condition in which the openings of the bile ducts and pancreatic ducts narrow.

Parietal cells - cells in the stomach wall that make hydrochloric acid.

Parkinsonism - the name given to a group of disorders with similar features including four primary symptoms: tremor, rigidity, postural instability, and bradykinesia, which are the result of the loss of dopamine-producing brain cells.

Parosmia - any disease or perversion of the sense of smell, especially the subjective perception of odors that do not exist.

Partial colectomy - the removal of part of the large intestine.

Partial mastectomy - surgery to remove a portion of the breast.

Passive accessory intervertebral movements (PAIVMs) - investigation of accessory gliding movements occurring in a joint.

Passive physiological intervertebral movements (PPIVMs) - investigation of passive physiological movements, passively, to confirm restrictions seen on active movement testing.

Patellar tendonitis - inflammatory condition of the patellar ligament, usually due to overuse.

Pathologist - physician who identifies diseases by studying cells and tissues under a microscope.

Pathology - the study of diseases.

Ppeak flow meter - a portable, inexpensive, hand-held device used to measure how air flows from lungs in one "fast blast" to measure the ability to push air out of the lungs.

Peak torque - measure of isokinetic performance.

Pediatric dentist - a specialist in the field of dentistry dealing particularly with the oral health care of children, from infancy through the teenage years.

Pelvic exam - a physician examination of the uterus, vagina, ovaries, fallopian tubes, bladder, and rectum.

Pelvic lymph node dissection - removal of some lymph nodes from the pelvis.

Pepsin - enzyme made in the stomach which breaks down proteins.

Peptic - related to the stomach and the duodenum, where pepsin is present.

Peptic ulcer - sore in the lining of the esophagus, stomach, or duodenum, usually caused by the bacterium Helicobacter pylori. An ulcer in the stomach is a gastric ulcer; an ulcer in the duodenum is a duodenal ulcer.

Perception - process of knowing or being aware of information through the ear.

Perfusion - flow

Perianal - area around the anus.

Pericardium - the membrane which surrounds the heart.

Perilymph fistula - leakage of inner ear fluid to the middle ear that occurs without apparent cause or is associated with head trauma, physical exertion, or barotrauma.

Perineal - related to the perineum.

Perineum - area between the anus and the sex organs.

Periodontal diseases (also called gum diseases) - serious bacterial infections which destroy the gums and the surrounding tissues of the mouth.

Periodontist - a specialist in the field of dentistry responsible for the care and prevention of gum-related diseases, guided bone regeneration, and dental implants.

Periostitis (also called shin splints) - inflammatory condition of the covering of the bones.

Peripheral stem cell transplantation - a process in which the stem cells (immature cells from which blood cells develop) are removed, treated with anticancer drugs, and frozen until they are returned to the patient.

Peritoneal adhesions - the peritoneum is a two-layered membrane which lines the wall of the abdominal cavity and covers abdominal organs. Sometimes organs begin to adhere to the peritoneum, requiring surgery to free the organs again.

Peritoneum - lining of the abdominal cavity.

Pessary - rubber or plastic device which is inserted through the vagina to help hold the uterus in place in women who have prolapse of the uterus.

Phalen's test - test for carpal tunnel syndrome in which the wrists are flexed for one minute.

Pharynx - space behind the mouth that serves as a passage for food from the mouth to the esophagus and for air from the nose and mouth to the larynx.

Phonology - study of speech sounds.

Photophobia - sensitivity to light.

Photorefractive keratectomy (PRK) - surgical procedure using an excimer laser to change the shape of the cornea.

Physiatrist - the physician who specializes in physical medicine and rehabilitation.

Pituitary gland - gland at the base of the brain which secretes hormones and regulates and controls other hormone-secreting glands and many body processes, including reproduction.

Placenta - organ which develops in the uterus during pregnancy; links the blood supplies of a pregnant woman to the fetus to provide nutrients and remove waste products.

Placental abruption - premature detachment of the placenta from the wall of the uterus causing severe bleeding that is life-threatening to both a pregnant woman and fetus.

Plantar warts - warts that occur on the sole of the foot and look like calluses; often a result of an infection or a specific virus.

Plaque, dental - film of mucus which traps bacteria on the surface of the teeth.

Plasma - the watery, straw-colored fluid which carries the cellular elements of the blood through circulation.

Platelets - cells found in the blood.

Pneumonectomy - removal of an entire lung, for cancer, lung abscesses, bronchiectasis, or extensive tuberculosis.

Polyp - a growth that projects from the lining of mucous membrane, such as the intestine.

Polyposis - presence of many polyps.

Polyunsaturated fat - a type of fat found in vegetable oils and margarines which does not appear to raise blood cholesterol levels.

Pontic tooth - false tooth.

Porcelain veneers - a ceramic material that is bonded to the front of teeth to change the tooth's color, size, and/or shape.

Portal hypertension - abnormally high blood pressure in the portal vein, which supplies the liver with blood from the intestine.

Portal vein - large vein which carries blood from the intestines and spleen to the liver.

Portosystemic shunt - operation to create an opening between the portal vein and other veins around the liver.

Positron emission tomography (PET) - a nuclear scanning procedure which gives a three-dimensional picture of the heart to provide information about the flow of blood through the coronary arteries to the heart muscle.

Post-anesthesia care unit (also called recovery room) - the area a patient is brought to after surgery to recover.

Postcholecystectomy syndrome (also called biliary dyskinesia) - condition which occurs after a gallbladder removal in which the muscle between the gallbladder and the small intestine does not work properly, causing pain, nausea, and indigestion.

Posterior chamber of the eye - the back section of the eye's interior.

Posterior optical segment - portion of the eye located behind the crystalline lens; includes the vitreous, choroid retina, and optic nerve.

Posterior vitreous detachment (PVD) - the separation of the vitreous from the retina.

Postgastrectomy syndrome - condition which occurs after an operation to remove the stomach (gastrectomy).

Postlingually deafened - individual who becomes deaf after having learned language.

Postvagotomy stasis - delayed stomach emptying; often occurs after surgery on the vagus nerve.

Predisposition - tendency to develop a certain disease.

Prednisolone - one type of corticosteroid medication, usually used for inflammation.

Prelingually deafened - individual who is either born deaf or who has lost hearing early in childhood, before learning language.

Presbycusis - loss of hearing which gradually occurs because of changes in the inner or middle ear in individuals as they grow older.

Preterm labor - labor that begins before the 37th week of pregnancy.

Prevalence - statistic that equals the total number of people in a population with a certain disease at a given time.

Prick skin test - a test to determine if a person is allergic to certain substances. A physician places a drop of the substance being tested on the persons' forearm or back and pricks the skin with a needle, allowing a tiny amount to enter the skin.

Primary sclerosing cholangitis - inflammation, scarring, and narrowing of the bile ducts inside and outside the liver.

Proctalgia fugax - intense pain in the rectum that occasionally happens at night; caused by muscle spasms around the anus.

Proctectomy - operation to remove the rectum.

Proctocolectomy (also called coloproctectomy) - operation to remove the colon and rectum.

Proctocolitis - inflammation of the colon and rectum.

Proctologist - physician who specializes in disorders of the anus and rectum.

Proctoscope - short, rigid metal tube used to look into the rectum and anus.

Proctoscopy - looking into the rectum and anus with a proctoscope.

Proctosigmoiditis - inflammation of the rectum and the sigmoid colon.

Proctosigmoidoscopy - endoscopic examination of the rectum and sigmoid colon.

Progestin - synthetic form of the female sex hormone progesterone.

Prognosis - a prediction of the course of the disease.

Prolapse - condition which occurs when a body part slips from its normal position.

Prolapse of the uterus - displacement of the uterus down into the vagina; caused by a weakening of supporting tissues in the pelvis.

Proprioceptive neuromuscular facilitation (PNF) - techniques used to improve motor skills through positive motor transfer, using the principles of facilitation / inhibition; irradiation / reinforcement; reciprocal innervation.

Prostatalgia - pain in the prostate gland.

Prostate acid phosphatase (PAP) - an enzyme produced by the prostate which is elevated in some patients when prostate cancer has spread beyond the prostate.

Prostatectomy - surgical procedure for the partial or complete removal of the prostate.

Prostate-specific antigen (PSA) - an antigen made by the prostate gland and found in the blood; may indicate cancer in the prostate gland.

Prostate-specific antigen (PSA) blood test - a blood test used to help detect prostate cancer by measuring a substance called prostate-specific antigen produced by the prostate.

Prostatism - any condition of the prostate that causes interference with the flow of urine from the bladder.

Prostatitis - an inflamed condition of the prostate gland which may be accompanied by discomfort, pain, frequent urination, infrequent urination, and, sometimes, fever.

Prosthodontist - a dental specialist who has undergone additional training and certification in the restoration and replacement of broken teeth with crowns, bridges, or removable prosthetics (dentures).

Proton pump inhibitors - medications which stop the stomach's acid pump.

Pruritus ani - itching around the anus.

Pubic symphysis - anterior joint of the pelvis.

Pudendal block - pain-relieving procedure used during childbirth in which an anesthetic is injected into tissues surrounding the pudendal nerves on either side of the vagina; blocks pain in the tissues between the vagina and anus.

Pulmonary - pertains to lungs and respiratory system.

Pulmonary artery - blood vessel delivering oxygen-poor blood from the right ventricle to the lungs.

Pulmonary hypertension - abnormally high blood pressure in the arteries of the lungs.

Pulmonary vein - the vessel which carries newly oxygenated blood to the heart from the lungs.

Pupil - the dark center in the middle of the iris through which light passes to the back of the eye.

Pupillary response - the constriction or dilation of the pupil as stimulated by light.

Pyloric sphincter - muscle between the stomach and the small intestine.

Pylorus - opening from the stomach into the top of the small intestine (duodenum).

Pyramidal pathway - a collection of nerve tracts which travel from the cerebral cortex through the pyramid of the medulla oblongata in the brainstem to the spinal cord. Within the pyramid of the medulla, fibers cross from one side of the brain to the opposite side.

Q-angle - direction of the quadriceps.

Qi - Chinese word for life force.

Quadriceps - large muscles located on the front of each thigh.

Radial keratotomy - a surgical procedure in which incisions are made into the epithelium of the cornea to correct refractive error.

Radiation - use of high-energy radiation from x-rays, neutrons, and other sources to kill cancer cells and shrink tumors.

Radiation colitis - inflammation of the colon from radiation therapy.

Radical mastectomy - surgical removal of the entire breast, the pectoral muscles, and the ancillary lymph nodes.

Radical prostatectomy - surgery to remove the prostate along with the two seminal vesicle glands attached to the prostate.

Radical retropubic prostatectomy - an operation to remove the entire prostate gland and seminal vesicles through the lower abdomen.

Radiculopathy - pinched nerve usually from a herniated or slipped disk; can cause a shooting pain often described as an electrical feeling.

Radioisotope - a radioactive material injected into the body so that a nuclear scanner can make pictures.

Radionuclide scan - an imaging scan in which a small amount of radioactive substance is injected into the vein. A machine measures levels of radioactivity in certain organs, thereby detecting any abnormal areas or tumors.

Radionuclide ventriculography - a diagnostic procedure used to determine the shape and size of the heart's chambers.

Radiopharmaceutical (also called tracer or radionuclide) - the basic, radioactively-tagged compound necessary to produce a nuclear medicine image.

Range of motion - measurement of the extent to which a joint can go through all of its normal range of movements.

Rapid gastric emptying - see dumping syndrome.

Recommended dietary allowance (RDA) - recommendations for daily intake of specific nutrients for groups of healthy individuals, as set by the Food and Nutrition Board of the National Research Council of the National Academy of Science.

Rectal manometry - test which uses a thin tube and balloon to measure pressure and movements of the rectal and anal sphincter muscles.

Rectocele - condition in which weakening of the lower vaginal wall causes the rectum to bulge into the vagina.

Rectum - lower end of the large intestine, leading to the anus.

Reflex sympathetic dystrophy (RSD) - caused by an abnormal sympathetic muscle reflex, giving rise to a response that is out of proportion to, and inconsistent with, the injury.

Reflux (also called regurgitation, GSR) - condition which occurs when gastric juices, or small amounts of food from the stomach, flow back into the esophagus and mouth.

Reflux esophagitis - irritation of the esophagus due to stomach contents flowing back into the esophagus.

Refractive error - the degree to which light reaches the back of the eye; includes myopia, hyperopia, and astigmatism.

Regional anesthetic - an anesthetic used to numb a portion of the body.

Regurgitation, gastrointestinal - see Reflux.

Regurgitation, heart - backward flow of blood caused by a defective heart valve.

Rehabilitation - the process of restoring a part of the body or a person to near-normal functioning after an injury or disease.

Required surgery - an operation which is necessary to continue quality of life. Required surgery may not have to be done immediately, like emergency surgery.

Respiration - gas exchange from air to the blood and from the blood to the body cells.

Respiratory system - the group of organs responsible for carrying oxygen from the air to the bloodstream and for expelling carbon dioxide.

Resting tremor - a tremor of a limb which increases when the limb is at rest.

Retching - dry vomiting.

Retrolisthesis - posterior slippage of one vertebra onto another.

Retropulsion - the tendency to step backwards if bumped from the front, or upon starting to walk; usually seen in patients who tend to lean backwards because of problems with balance.

Rheumatic fever - a childhood disease which may damage the heart valves or the outer lining of the heart.

Rhinitis - an inflammation of the mucous membrane which lines the nose; often due to allergy to pollen, dust, or other airborne substances; causes sneezing, itching, a runny nose, and nasal congestion.

RICE (rest, ice, compression, and elevation) - treatment plan for acute injury to prevent inflammatory processes from becoming uncontrolled, and to speed up the recovery process by eliminating swelling; acute injury management.

Rigidity - increased resistance to the passive movement of a limb.
Risk factor - activity or factor which may increase the chance of developing a disease.

Rotator cuff - muscles and their insertional tendons that form a cuff over the shoulder joint, on their way to attaching from the scapula to the humorous; major function is to control, and produce, rotation of the shoulder.

Round window - membrane separating the middle ear and inner ear.

Rupture - break or tear in any organ or soft tissue.

Sacro-iliac joint - synovial joint formed between the medial surface of the ilium and the lateral aspect of the upper sacral vertebrae; a fibrous joint, comprising and surrounded by very strong ligamentous structures; major function is to transmit body weight.

Saline solution - a solution containing sodium chloride.

Saliva - mixture of water, protein, and salts which makes food easy to swallow; begins digestion.

Salpingectomy - surgical removal of one or both fallopian tubes.

Salpingo-oophorectomy - surgery to remove the fallopian tubes and ovaries.

Saturated fat - fat which is found in foods from animal meats and skin, dairy products, and some vegetables.

Scapula - commonly known as the shoulder blade.

Sciatica - traditionally defined as pain in the sciatic nerve, or its branches, which is caused by direct pressure or irritation.

Scintigraphy - see radionuclide scans.

Sclerotherapy - method of stopping upper gastrointestinal bleeding. A needle is inserted through an endoscope to bring hardening agents to the place that is bleeding.

Scotoma - an area of partial or complete loss of vision surrounded by an area of normal vision.

Sebum - oily substance produced by sebaceous glands in the skin.

Secretin - hormone made in the duodenum; causes the stomach to make pepsin, the liver to make bile, and the pancreas to make digestive juices.

Segmental mastectomy - surgery to remove a portion of the breast.

Self-monitoring blood glucose - method for persons with diabetes to find out how much glucose is in their blood, usually obtained by pricking a finger for a small drop of blood.

Sella turcica - bony structure which houses the pituitary gland.

Sensorineural hearing loss - hearing loss caused by damage to the sensory cells and/or nerve fibers of the inner ear.

Septal defect - a hole in the wall of the heart.

Septum - the muscle wall which divides the heart chambers.

Serum - a clear fluid which separates when blood clots.

Sesamoiditis - inflammatory condition affecting the two sesamoid bones in the flexor tendon of the great (big) toe.

Sexually transmitted disease (STD) - infection spread through sexual intercourse and other intimate sexual contact.

Shigellosis - infection with the bacterium Shigella; usually causes a high fever, acute diarrhea, and dehydration.

Short gut syndrome - see short bowel syndrome.

Shunt - a connector which allows blood flow between two locations.

Shwachman's syndrome - digestive and respiratory disorder of children in which certain digestive enzymes are missing and white blood cells are few.

Sialorrhea - drooling.

Sigmoid colon - lower part of the colon that empties into the rectum.

Sign language - language of hand shapes, facial expressions, and movements used as a form of communication, mainly by hearing impaired persons.

Silent ischemia - ischemia not accompanied by chest pain.

Simple mastectomy - surgical removal of the breast and possibly a few of the axillary lymph nodes close to the breast.

Sinus (also called paranasal sinuses) - air cavities within the facial bones, lined by mucous membranes similar to those in other parts of the airways.

Sinus node - the cells which produce the electrical impulses that cause the heart to contract.

Slump test - a test for assessing the mobility of the pain-sensitive structures in the vertebral canal.

Small bowel enema - x-rays of the small intestine taken as barium liquid passes through the organ. (see also lower GI series)

Small intestine - organ where most digestion occurs; measures about 20 feet long, and includes the duodenum, jejunum, and ileum.

Smell - to perceive odor or scent through stimuli affecting the olfactory nerves.

Smell disorder - temporary or permanent inability to perceive odors.

Smooth muscle - muscle which performs automatic tasks, such as constricting blood vessels.

Soft tissue - generally, the ligaments, tendons, and muscles in the musculoskeletal system.

Somatosensory - refers to sensory signals from all tissues of the body including skin, viscera, muscles, and joints.

Somatostatin - a hormone in the pancreas which helps tell the body when to make the hormones insulin, glucagon, gastrin, secretin, and renin.

Spasm - a condition in which a muscle or group of muscles involuntarily contract.

Spasmodic dysphonia - momentary disruption of the voice caused by involuntary movements of one or more muscles of the larynx (or voice box).

Spastic colon - see irritable bowel syndrome (IBS).

Specific language impairment (SLI) - difficulty with the organized symbol-system communication in the absence of problems such as mental retardation, hearing loss, or emotional disorders.

Speech - making definite vocal sounds which form words to express thoughts and ideas.

Speech processor - part of a cochlear implant which converts speech sounds into electrical impulses to stimulate the auditory nerve.

Speech-language pathologist - health professional trained to evaluate and treat people who have voice, speech, language, or swallowing disorders, including hearing impairments, that affect their ability to communicate.

Sphincter - ring-like band of muscle which opens and closes an opening in the body.

Sphincter of Oddi - muscle between the common bile duct and pancreatic ducts.

Sphygmomanometer - the instrument used to measure blood pressure.

Spinal anesthetic (also known as Spinal tap) - an anesthetic which is injected into the spinal canal fluid for surgery in the lower abdomen, pelvis, rectum, or other lower extremities.

Spinal instability - increased motion between vertebra, usually resulting from an injury; pain typically feels like tingling in the neck or arms.

Spine - a column in the body consisting of 33 vertebrae.

Spirogram - record of the amounts of air being moved in and out of the lungs.

Spleen - organ lying between the stomach and diaphragm which stores red blood cells and filters blood.

Splenectomy - surgical removal of the spleen.

Splenic flexure syndrome - condition which occurs when air or gas collects in the upper parts of the colon.

Spondylosis - a condition of the spine involving a degenerative process.

Sputum (also called phlegm) - mucous from the lungs.

Stage - the measurement of the extent of a cancer or other disease, whether the disease has spread from the original site to other parts of the body, or the disease has progressed.

Staging - an evaluation of the extent of disease which provides the basis for making treatment recommendations.

Steatorrhea - condition in which the body cannot absorb fat.

Steatosis - see fatty liver.

Stenosis - the narrowing or constriction of a blood vessel or valve in the heart or other organ.

Stereopsis - ability to perceive three-dimensional depth.

Sternum - the breastbone.

Stethoscope - the instrument used to listen to the heart and other sounds in the body.

Stirrups - technique of ankle strapping using rigid tape placed on the ankle; medial to lateral adhering to the undersurface of the heel, mimicking a stirrup.

Stoma - opening in the abdomen which is created by an operation (ostomy); must be covered at all times by a bag that collects stool. (reference Ostomy)

Stomach - main organ of digestion which lies between the esophagus and the small intestine.

Stool (also called feces) - solid wastes which pass through the rectum as bowel movements. Stools are undigested foods, bacteria, mucus, and dead cells.

Straight leg raise (SLR) - technique for measuring sciatic nerve mobility and/or hamstring length.

Streptokinase - a clot-dissolving medication.

Stress - mental or physical tension which results from physical, emotional, or chemical causes.

Stress fracture - fracture caused by non-traumatic, cumulative overload on a bone; caused by a number of factors including overtraining, incorrect biomechanics, fatigue, hormonal imbalance, poor nutrition, and osteoporosis.

Stress incontinence - involuntary leaking of urine which may be caused by coughing or straining.

Stress ulcer - upper gastrointestinal ulcer from physical injury such as surgery, major burns, or critical head injury.

Striatum - part of the basal ganglia; a large cluster of nerve cells, consisting of the caudate nucleus and the putamen, which controls movement, balance, and walking; the neurons of the striatum require dopamine to function.

Stricture (also called stenosis) - abnormal narrowing of a body opening.

Subacromial joint - a joint made up by the humorous and a superior arch, consisting of the acromion process and the coracoid process of the scapula; joined by the coraco-acromial ligament and lined by the synovium of the subacromial bursa.

Subacromial space - region of the shoulder that is bordered by the subacromial joint.

Subchondral tissue - the smooth tissue at the ends of bones, which is covered with another type of tissue.

Substantia nigra - a small area of the brain containing a cluster of black-pigmented nerve cells which produce dopamine, which is then transmitted to the striatum.

Sudden deafness - loss of hearing which occurs quickly from such causes as explosion, a viral infection, or the use of some drugs.

Sudden death - death which occurs unexpectedly or immediately after onset of symptoms.

Sudeck's atrophy - condition characterized by severe pain, swelling, and disability; commonly seen in the foot, but may occur in other joints of the lower limbs.

Sun Protection Factor (SPF) - Superior vena cava - the large vein which returns blood to the heart from the head and arms.

Suprarenal glands - another name for the adrenal glands.

Surgery - operation to remove or repair a part of the body, or to find out if disease is present.

Suspensory ligament of lens - a series of fibers which connect the ciliary body of the eye with the lens, holding it in place.

Sustention tremor (also called postural tremor) - a tremor of a limb that increases when the limb is stretched.

Synapse - a tiny gap between the ends of nerve fibers across which nerve impulses pass from one neuron to another; at the synapse, an impulse causes the release of a neurotransmitter, which diffuses across the gap and triggers an electrical impulse.

Syncope - light-headedness or fainting caused by insufficient blood supply to the brain.

Syndromic hearing impairment - hearing loss or deafness which is inherited or passed through generations of a family.

Synovitis - inflammation of the synovial membrane, the tissue that lines and protects the joint.

Synovium - a fibrous envelope which produces a fluid that helps to reduce friction and wear in a joint.

Systemic treatment (also called systemic therapy) - treatment or therapy which reaches and affects cells throughout the body.

Systolic blood pressure - the highest pressure to which blood pressure rises with the contraction of the heart.

Tachycardia - rapid heart beat.

Tachypnea - rapid breathing.

Tamoxifen - an anticancer drug used in hormone therapy to block the effects of estrogen.

Taste - sensation produced by a stimulus applied to the gustatory nerve endings in the tongue; the four tastes include: salt, sour, sweet, and bitter; some say there is a fifth taste described as savory.

Taste buds - groups of cells located on the tongue which enable one to recognize different tastes.

Taste disorder - inability to perceive different flavors.

Technetium bone scan - radiological investigation to detect stress fractures.

Telemetry unit - a small transmitter with wires that attach ECG patches to the chest which is used to send information about the heart, via radio transmission, to healthcare professionals for evaluation.

Temporal arteries - vessels located over the temples on each side of the head which supply blood to part of the head.

Tendon - a cord which connects muscle to bone or other tissue.

Tendonitis - inflammation of a tendon or of the tendon covering; usually caused by not just a single injury, but a series of small stresses that repeatedly aggravate the tendon.

Thallium stress test - a study in which radioactive potassium is carried by the blood; the progress is followed by x-ray pictures.

Thomas's test - examination to detect a fixed hip deformity.

Thoracotomy - surgery to view the lung (that may be used to confirm cancer), or for chest trauma (to detect the source of bleeding).

Throat disorders - disorders or diseases of the larynx (voice box) or esophagus.

Thrombolysis - the breaking up of a blood clot.

Thrombolytic drugs - medications used to dissolve blood clots.

Thrombosis - an abnormal condition in which a blood clot develops in the blood vessel or in the heart.

Thrombosis, deep-vein - formation of blood clots in veins deep inside the legs.

Thrombus - a blood clot.

Thyroplasty (also called laryngeal framework surgery) - surgical technique to improve the voice by altering the cartilages of the larynx.

Thyroxin (T4) - a hormone secreted by the thyroid gland which regulates metabolism.

Tibia - shin bone.

Tinnitus - sensation of a ringing, roaring, or buzzing sound in the ears or head; often associated with various forms of a hearing impairment.

Tissue - group or layer of cells which together perform specific functions.

Tissue plasminogen activator (TPA) - a medication used to dissolve blood clots.

Tomography - from the Greek words "to cut" or "section" (tomos) and "to write" (graphein); in nuclear medicine, it is a method of separating interference from the area of interest by imaging a cut section of the object.

Tongue - large muscle on the floor of the mouth which manipulates food for chewing and swallowing; the main organ of taste; assists in forming speech sounds.

Total gastrectomy - complete removal of the stomach.

Total hysterectomy - the removal of the uterus, including the cervix; the fallopian tubes and the ovaries remain.

Total hysterectomy with bilateral salpingo-oophorectomy - the entire uterus, fallopian tubes, and the ovaries are surgically removed.

Tourette syndrome - neurological disorder characterized by recurring movements and sounds (called tics).

Tracheoesophageal fistula (TEF) - condition which occurs when there is a gap between the upper and lower segments of the esophagus, through which food and saliva cannot pass.

Transcutaneous electrical nerve stimulation (TENS) - a method of providing pain relief using electrical signals which are sent to the nerve endings.

Transesophageal echocardiography (TEE) - a diagnostic test which is used to measure the sound waves that bounce off of the heart.

Transient ischemic attack (TIA) - a stroke-like event that lasts for a short period of time, caused by a blocked blood vessel.

Transplantation - replacing a damaged organ with one from a donor.

Transrectal ultrasound of the prostate - a test using sound wave echoes to create an image of the prostate gland to visually inspect for abnormal conditions such as gland enlargement, nodules, penetration of tumor through capsule of the gland, and/or invasion of seminal vesicles;

Transurethral incision of the prostate (TUIP) - a procedure which widens the urethra by making some small cuts in the bladder neck, where the urethra joins the bladder, and in the prostate gland itself.

Transurethral resection of the prostate (TURP) - surgery to remove part of the prostate gland which surrounds the urethra by using a small tool that is placed inside the prostate through the urethra; no incision is made using this method.

Transurethral surgery - surgery in which no external incision is needed; the surgeon reaches the prostate by inserting an instrument through the urethra.

Transvaginal ultrasound (also called ultrasonography) - an ultrasound test using a small instrument, called a transducer, which is placed in the vagina.

Transverse colon - part of the colon which extends across the abdomen from right to left.

Trendelenburg gait - intrinsic disorder of the abductors of the hip, due to either a weakness or an inhibition to function; hip abductors are unable to stabilize the hip, as body weight is transferred to the affected side, resulting in a pelvic drop.

Tricuspid valve - the heart valve which controls blood flow from the right atrium into the right ventricle.

Trigger finger - an irritation of the digital sheath that surrounds the flexor tendons of the finger. When the tendon sheath becomes thickened or swollen, it pinches the tendon and prevents it from gliding smoothly. In some cases, the tendon catches and then suddenly releases.

Trigger point - hypersensitive area or site in muscle or connective tissue; usually associated with myofascial pain syndromes.

Trimester - a period of three months.

Tube feeding - see Enteral nutrition.

Type 1 diabetes (also called insulin dependent diabetes) - a condition in which the pancreas produces so little insulin that the body cannot use blood glucose as energy; must be controlled with daily insulin injections.

Type 2 diabetes (also called non-insulin dependent diabetes and adult-onset diabetes) - a condition in which the pancreas produces less insulin than needed so that the body cannot use the blood glucose as energy properly; can often be controlled through meal plans and physical activity plans, and diabetes pills or insulin.

Tyrosine - the amino acid from which dopamine is made.

Ulcer - sore on the skin surface or on the stomach lining.

Ultraviolet radiation - invisible rays which come from the sun. UV radiation can damage the skin and cause melanoma and other types of skin cancer.

Upper GI endoscopy - examining into the esophagus, stomach, and duodenum with an endoscope.

Upper limb tension test (ULTT) - test to assess pain.

Urea breath test - test used to detect Helicobacter pylori infection. The test measures breath samples for urease, an enzyme H. pylori produces.

Ureters - two tubes which carry urine from the kidneys to the bladder.

Urethra - narrow channel through which urine passes from the bladder out of the body.

Urge incontinence - the inability to hold urine long enough to reach a restroom.

Urgent surgery - an operation performed immediately as a result of an urgent medical condition.

Urinary retention - the inability to empty the bladder.

Uro, urono - when used as a part of a word, refers to the urinary tract.

Usher's syndrome - a hereditary disease which affects hearing and vision.

Uterus (also called the womb) - the uterus is a hollow, pear-shaped organ located in a woman's lower abdomen, between the bladder and the rectum, which sheds its lining each month during menstruation, and in which a fertilized egg implants and grows into a fetus.

Vaccine: a substance which has been created from weakened or dead bacteria or viruses that is used to stimulate the body's immune defense to that disease or virus.

Vaginal hysterectomy - the uterus is removed through the vaginal opening.

Vaginosis, bacterial - very common vaginal infection characterized by symptoms such as increased vaginal discharge or itching, burning, or redness in the genital area.

Vagotomy - operation to cut the vagus nerve, which causes the stomach to produce less acid.

Vagus nerve - nerve in the stomach which controls the production of stomach acid.

Valgus deformity - a lateral inclination of a distal bone of a joint from the midline.

Valve - fold in the lining of an organ which prevents fluid from flowing backward.

Valves, heart - (the heart valves are tricuspid, pulmonary, mitral, and aortic) - the "doors" between the chambers of the heart.

Valvuloplasty - repairing a heart valve.

Varices - stretched veins such as those which form in the esophagus from cirrhosis.

Varicose vein - an abnormally dilated vein, usually found in the legs.

Varus deformity - a medial inclination of a distal bone of a joint from the midline.

Vascular - pertaining to blood vessels.

Vasodepressors - a medication which raises blood pressure.

Vasodilator - agent which widens blood vessels.

Vein - a blood vessel which carries blood from the body back into the heart.

Velocardiofacial syndrome - an inherited disorder characterized by cleft palate, heart defects, a characteristic facial appearance, minor learning problems, and speech and feeding problems.

Ventilation - movement of air (gases) in and out of the lungs.

Ventricle - one of the two pumping chambers of the heart; the right ventricle receives oxygen-poor blood from the right atrium and pumps it to the lungs through the pulmonary artery; the left ventricle receives oxygen-rich blood from the left atrium.

Ventricular fibrillation - a condition in which the ventricles contract in rapid and unsynchronized rhythms and cannot pump blood into the body.

Vertebral artery testing - evaluation procedure to assess the relationship between cervical spine movement and symptoms which may be vertebrobasilar in origin.

Vestibular neuronitis - infection at the vestibular nerve.

Vestibular system - system in the body which is responsible for maintaining the body's orientation in space, balance, and posture; regulates locomotion and other movements and keeps objects in visual focus as the body moves.

Vestibule - bony cavity of the inner ear.

Vibrotactile aids - mechanical instruments which help individuals who are deaf detect and interpret sound through the sense of touch.

Villi - tiny, finger-like projections on the surface of the small intestine which help absorb nutrients.

Viral hepatitis - hepatitis caused by a virus. Five different viruses (A, B, C, D, and E) most commonly cause this form of hepatitis. Other rare viruses may also cause hepatitis.

Vitrectomy - an operation to remove the blood which sometimes collects at the back of the eyes when a person has eye disease.

Vitreous body - a clear, jelly-like substance that fills the center of the eye.

Virus - a microorganism which can live only by attaching itself to a living cell, and which can cause diseases.

Vocal cord paralysis - inability of one or both vocal folds (vocal cords) to move because of damage to the brain or nerves.

Vocal cords (also called vocal folds) - muscularized folds of mucous membrane which extend from the larynx (voice box) wall; enclosed in elastic vocal ligament and muscle which control the tension and rate of vibration of the cords as air passes through them.

Vocal tremor - trembling or shaking of one or more of the muscles of the larynx - resulting in an unsteady-sounding voice.

Voice - sound produced by air passing out through the larynx and upper respiratory tract.

Voice disorders - group of problems involving abnormal pitch, loudness, or quality of the sound produced by the larynx (voice box).

Volvulus - twisting of the stomach or large intestine.

Vomiting - release of stomach contents through the mouth.

Watchful waiting (also called expectant management) - close monitoring of cancer by a physician instead of immediate treatment.

Watermelon stomach - parallel red sores in the stomach which look like the stripes on a watermelon.

Wedge resection of the lung - a small, localized section of the lung is removed, often for a lung biopsy.

Whole blood - blood containing all its components, such as red and white blood cells, platelets, and more.

Yin and Yang - Chinese words for complementary and opposite forces that make up the life force (Qi).

Yoga - Indian word for "union;" a posturing and breathing technique to induce relaxation.

ABOUT THE AUTHOR:

Dana Jacoby is the founder and president of Med-Ed, a consulting and medical education company focused on patient care and medical practices. Dana has over a decade of experience in healthcare, sales, and management, with a focus on specialties of medicine including Cardiology, Plastic Surgery, Neurology, Orthopedics, Dermatology, Oncology, and Internal Medicine. Dana's true passion throughout her tenure as a medical marketing and sales professional has always been patient and medical staff education. She has developed numerous sales training programs, marketing campaigns, and patient education seminars to further this cause.

Dana is a graduate of Louisiana State University and holds a Masters of Management from Tulane University School of Business. Additionally, she is a Certified Medical Professional through the CMR institute, a personal fitness and nutrition trainer through the International Sports Sciences Association, and is in the process of completing her Master of Health Systems from the University of Medicine and Dentistry in New Jersey.

An avid marathon runner, Dana's favorite quote is that any challenge in life can be "won" one step at a time.

May this book serve to assist with Dana's life-long dream of educating the nation one patient at a time.

Made in United States
Orlando, FL
04 November 2026

90205151R00127